The Wooing of Earth

ALSO BY RENÉ DUBOS

Louis Pasteur
The Dreams of Reason
Man Adapting
So Human an Animal
The Mirage of Health
A God Within
Beast or Angel?
The Professor, the Institute, and DNA
Chercher

THE WOOING OF EARTH

René Dubos

New York

CHARLES SCRIBNER'S SONS

Library of Congress Cataloging in Publication Data

Dubos, René Jules, 1901–
The wooing of Earth.

Includes index.
1. Nature conservation. 2. Landscape protection.
3. Man—Influence on nature. I. Title.
QH75.D8 304.2'8 79-28097
ISBN 0-684-16501-5

This book published simultaneously in the
United States of America and in Canada –
Copyright under the Berne Convention

1 3 5 7 9 11 13 15 17 19 F/P 20 18 16 14 12 10 8 6 4 2

Printed in the United States of America

To the billions of human beings
who, for thousands of years,
have transformed the surface of the Earth
and prepared it for civilization.

ACKNOWLEDGMENTS

MANY OF THE FACTS and ideas presented in the following pages have been derived (but not exactly reproduced) from three different sources: the column that I write regularly for *The American Scholar* under the title "The Despairing Optimist"; a lecture that I delivered at the University of Colorado in Boulder at the invitation of Professor Gilbert White and that was published in an extended form as a bulletin by the Colorado University Press under the title "The Resiliency of Ecosystems"; several unpublished lectures before societies of architecture and of landscape planning. I wish to thank the officers of the respective organizations for having given me the opportunity to formulate and document the general scheme I had in mind when I first proposed the title *The Wooing of Earth* to Charles Scribner's Sons more than seven years ago.

I am also grateful to the National Endowment for the Humanities and to several corporations for having financed during the past two years The René Dubos Forum, which has indirectly facilitated the writing of this book by organizing scholarly seminars and educational programs focused on the protection, management, and uses of humanized environments.

Finally, I want to thank my wife Jean and my associate Dr. Carol Moberg, both of whom have so richly contributed to the substance and form of *The Wooing of Earth* that I can recognize their presence in almost every page of the book.

CONTENTS

I remember how in my youth, in the course of a railway journey across Europe from Brindisi to Calais, I watched with keen delight and wonder that continent flowing with richness under the age-long attention of her chivalrous lover, western humanity . . . the heroic love-adventure of the West, the active wooing of the earth.

Rabindranath Tagore
Towards Universal Man

*Sans doute un jour, devant les étendues arides ou reconquises par la forêt, nul ne devinera plus ce que l'homme avait imposé d'intelligence aux formes de la terre en dressant les pierres de Florence dans le grand balancement des oliviers toscans.**

André Malraux
Les Voix du Silence

*Some day, facing a world become arid or once more overtaken by the trees, no one will realize how much intelligence man added to the forms of the earth, by erecting the monuments of Florence amidst the vast motions of the olive trees of Tuscany.—Translation by the author

PREFACE

On a sunny morning of June 1964, my wife and I were in the Boboli Gardens, admiring the city of Florence across the Arno River. It was a holy day and at noon, an immense paean of bells reached us from all the city's churches, each bell with its own voice. We were in an enchanted world, but the enchantment was purely of human creation. The whole Arno valley was a forested wilderness before historical times, whereas every part of it is now imbued with human intelligence. There are, of course, the monuments of Florence mentioned by Malraux in the phrase quoted, but also the olive trees that ancient people introduced into Tuscany from Asia. The Tuscan landscape has been sculpted and embellished by generations of peasants, prelates, and princes.

My first intention in writing this book was to deal exclusively with the creative and appealing aspects of human interventions into nature—the equivalents of the Tuscan countryside and its olive groves that can be found in many parts of the world, including the United States. Such a one-sided view of the environmental problem, however, might give the impression that I am blind to the present realities of life. I could not forget in any case that Florence suffered a disastrous flood in 1966, almost certainly caused by excessive deforestation of the Arno valley and rendered more destructive by the enormous amounts of oil carried by torrential waters from central heating plants. The world's environmental crisis has been discussed to death and hardly needs further elaboration, but it raised its ugly head whenever I considered the creative human interventions into natural systems. I could not celebrate the wooing of the Earth without constantly having in mind the rape of the Earth.

The Wooing of Earth thus turned out to be both sweet and sour. The sweet comes from my belief that human beings can improve on nature and from my knowledge that they can correct environmental damage by deliberate social action. The sour also has two ingredients: our propensity to spoil desirable environments, whether of natural or human origin, and my fear that nature's mechanisms of recovery may eventually fail to cope with our increasing use and misuse of resources and energy.

If sweet prevails over sour in my text, as I hope it does, it may simply be due to the fact that I have been happily conditioned by the very humanized environments of the Île de France, in which I was reared, and of the Hudson Highlands, where we own a once abandoned farm on which we are managing, with some success, a civilized return of the forest. I still regard these regions as more appealing now than they would be if they had remained in the state of wilderness, and I tend to regard in a similarly favorable way many other humanized environments all over the world. I admire the high Alps, the Sierras, and the Himalayas, but my love extends to less grandiose sceneries structured by human intelligence and imbued with the human presence. Certain areas of wilderness have been preserved (although not adequately protected), especially in North America, but it is important also to preserve areas where the partnership between humankind and Earth has generated values that transcend those created by natural forces working alone. A village green, a pasture, or a wheat field created by clearing the land have social, aesthetic, and emotional values of their own different from those of the surrounding wilderness.

In his poem "The Aim was Song" Robert Frost wrote of the wind as an "untaught" force of nature that "hadn't found the place to blow" until it had been redirected by the poet and converted into a song. More prosaically, I once used the French phrase "*La terre a besoin des hommes*" also to convey my belief that Earth has potentialities that remain unexpressed until properly manipulated by human labor and imagination. But work is not enough to discover and bring to light the hidden treasures of Earth; it also takes love. As Tagore wrote, the heroic love-adventure of humankind has been the wooing of Earth.

The Wooing of Earth

CHAPTER ONE

A Family of Landscapes

SOME of the landscapes that we most admire are the products of environmental degradation. The denuded islands of the Aegean Sea, the rocky shores of the Mediterranean basin, the semidesertic areas of the American Southwest are regions that appeal to countless people from all social and ethnic groups, as well as professional ecologists. Yet these landscapes derive much of their color and sculptural beauty from deforestation and erosion, the two cardinal sins of ecology. The immense majority of people, furthermore, elect to live in places from which the wilderness has been eradicated and which have been profoundly transformed by human habitation. Orthodox ecological criteria are therefore not adequate to evaluate the quality of a particular environment for human life.

Since the humanization of Earth inevitably results in destruction of the wilderness and of many living species that depend on it, there is a fundamental conflict between ecological doctrine and human cultures, a conflict whose manifestations are most glaring in Greece.

On two occasions during the past few years, I visited the

eleventh-century Byzantine monastery of Moni Kaisarianis, located some five miles southeast of Athens. The monastery is nestled on the slopes of Mount Hymettus at 1,100 feet elevation. A trail meanders from it toward the Hymettus mountain through an almost treeless landscape amidst thyme, lavender, sage, mint, and other aromatic plants. The rock formations of the area are denuded, but the luminous sky gives them an architectural quality particularly bewitching under the violet light of sunset.

A short distance from the monastery, the trail reaches an outcrop of rocks that affords a sudden view of the Acropolis, Mount Lycabettus, and the entire city of Athens. As is so often the case in Greece, the buildings—whether pagan or Christian—derive a dramatic quality independent of their architectural merit from their natural setting. But the landscape surrounding the monastery is not natural; it has been transformed by several thousand years of human occupation.

The grounds associated with the Moni Kaisarianis monastery are planted with almond and olive trees, two species that have long been part of the Greek flora but originated in south central or southeastern Asia. The road that leads from Athens to the monastery is shaded with eucalyptus trees introduced from Australia. Beyond the monastery, the Hymettus is stark and luminous but its rock formations were originally masked by earth and trees. Its bold architecture became clearly visible only during historical times as a result of deforestation and erosion.

Ecologists and historians agree that most of the Mediterranean world was wooded before human occupation. What we now regard as the typical Greek landscape, often stark and treeless, is the result of human activities. The rock structures were revealed only after the felling of the trees, which resulted in extensive erosion. The slopes have been kept denuded by rabbits, sheep, and goats that continuously destroy any new growth either of trees or grass. Erosion and overgrazing are the forces, inadvertently set in motion by human activities, that enable light to play its bewitching game on the white framework of Attica.

The humanization of the Greek wilderness has been achieved at great ecological loss. Writers of the classical, Hellenistic, and Roman periods were aware of the transformations brought about by deforestation in the Mediterranean world. In *Critias,* Plato compared the land of Attica to the "bones of a wasted body . . . the richer and softer parts of the soil having fallen away, and the mere skeleton being left." In ancient times, still according to Plato, the buildings had "roofs of timber cut from trees which were of a size sufficient to cover the largest houses." After deforestation, however, "the mountains only afforded sustenance to bees." The famous Hymettus honey is thus linked to deforestation, which permitted the growth of sun-loving aromatic plants.

As long as the mountain slopes were wooded, the land of Greece, as well as of other Mediterranean countries, was enriched by rainfall, but by Plato's time erosion caused the water to "flow off the bare earth into the sea. . . ." The sacred groves and other sanctuaries were originally established near springs and streams, but these progressively dried up as a consequence of deforestation.

The Ilissus River, which has its source on Mount Hymettus and runs through Athens, was still a lively stream in Plato's time. On a hot day in midsummer, Socrates and Phaedrus walked toward a tall plane tree on the banks of the Ilissus a short distance from the Agora. There, as reported in the famous dialogue, they discussed rhetoric, philosophy, and love while cooling their feet in the stream that they found "delightfully clear and bright." Today, the Ilissus is dry much of the year and, covered by a noisy roadway, serves as a sewer. There could not be a more dramatic symbol of the damage done by deforestation, erosion, and urban mismanagement.

The reforestation of Greece would certainly result in climatic and agricultural improvements. As Henry Miller writes in *The Colossus of Maroussi,* "The tree brings water, fodder, cattle, produce . . . shade, leisure, song. . . Greece does not need archeologists—she needs arboriculturists." But a cover of trees would make the landscape very different from the image that we, and the Greeks themselves, have had of Greece since classical times. In his poem "The Satyr or the Naked Song," the Greek poet Kostes Palamas (1859–1943) sees in the stark eroded structures of the present landscape a symbol of the austerity and purity of the Greek genius; the landscape triumphantly proclaims the "divine nudity" of Greece. Henry Miller himself, a few pages before and after the passage quoted above in which he advocates reforestation, marvels at the quality given to the landscape by the rocks that "have been lying for centuries exposed to this divine illumination . . . nestling amid dancing colored shrubs in a blood-stained soil." In Miller's words, these rocks "are symbols of life eternal." He does not mention that they are visible only because of deforestation and erosion.

While visiting the Moni Kaisarianis monastery, I noticed a dark opaque zone on the slopes of Mount Hymettus; this area had been reforested with pines. To me, it looked like an inkblot on the luminous landscape, especially at sunset, when the subtle violet atmosphere suffuses the bare rocks throughout the mountain range. The "divine illumination" lost much of its magic where it was absorbed by the pine trees.

The mountains of Attica were probably difficult to penetrate and frightening when completely wooded, but they have now acquired some of the qualities of a park. The traveler can move on their open surfaces, and vision can extend into a distance of golden light. I have wondered whether the dark and ferocious divinities of the preclassical Greek period did not become more serene and more playful precisely because they had emerged from the dark forests into the open landscape. Would logic have flourished if Greece had remained covered with an opaque tangle of trees?

There is no doubt that people spoiled the water economy and impoverished the land when they destroyed the forests of the Mediterranean world. But it is true also that deforestation allowed the landscape to express certain of its potentialities that had remained hidden under the dense vegetation. Not only did removal of the trees permit the growth of sun-loving aromatic plants and favor the spread of honeybees, as Plato had recognized; more importantly, it revealed the underlying architecture of the area and perhaps helped the soaring of the human mind.

The full expression of the Mediterranean genius may require both the cool mysterious fountains in the sacred groves and the bright light shining on the sun-loving plants amid the denuded rocks. Ecology becomes a more complex but far more interesting science when human aspirations are regarded as an integral part of the landscape.

THE wide range of natural and humanized environments of which we have knowledge calls to mind the collection of photographs that Edward Steichen published a few years ago under the title *The Family of Man.* In his book, Steichen created a panorama of the human species as it can be found today all over the Earth in its richly diversified types—the unloved and miserable as well as the loved and glorious, those conveying a sad resignation as well as those radiating a challenging beauty. Just as Steichen conceived his collection to be, in his own words, "a mirror of the universal elements and emotions in the everydayness of life," so I shall refer to many different aspects of the Earth, some that evolved under the influence of natural forces and others that have been completely transformed by human activities—whether these were ecologically constructive or destructive.

Many people reject Steichen's view that beauty can be found in all the visages of the family of man, and even more will be disturbed by my statement that deforestation and erosion have produced some of the most admired landscapes. It is obvious, of course, that the Earth and its atmosphere have been spoiled in many places by human carelessness and greed, but I feel nevertheless that almost any kind of scenery, even artificial and desolate, can be a source of interest and pleasure if one knows how to recognize in it patterns of visual organization, aspects of ecological curiosity, and matters of human concern.

I shall focus in the following pages on the interplay between human life and the various aspects of the Earth: the wilderness environments that are the undisturbed expressions of natural forces and the humanized environments that have emerged in the course of social evolution.

CHAPTER TWO

The Wilderness Experience

UNTIL the age of twenty-seven, I knew of the Earth only some of its most humanized environments, in France, Italy, England, and the eastern coast of the United States. My first direct contacts with the wilderness were during the late 1920s and the 1930s, when I drove several times across the North American continent and discovered—a true emotional discovery for me—the New Mexico mesas from the Raton Pass and the Pacific Ocean from a primeval forest in Oregon. Around 1930, these environments were still essentially wild.

I now realize how much my life would have been enriched by longer and more intimate contacts with the wilderness. The experience of nature in a native prairie, a desert, a primeval forest, or high mountains not crowded with tourists is qualitatively different from what it is in a well-tended meadow, a wheat field, an olive grove, or even in the high Alps. Humanized environments give us confidence because nature has been reduced to the human scale, but the wilderness in whatever form almost compels us to measure ourselves against the cosmos. It makes us realize how insignificant we are as biological creatures and invites us to escape from daily life into the realms of eternity and infinity.

In one of Kyoto's Zen temples, I have seen men and women who gave the impression of achieving this escape by looking at a distant hill in an attitude of reverence. We can also perceive some of the cosmic values of the wilderness by contemplating the great spectacles of nature, for example, simply by looking down into the Grand Canyon of the Colorado. But the real experience of the wilderness probably requires the participation of all our senses, in a manner that calls to mind Paleolithic ways of life.

The hunter-gatherers of the Old Stone Age were conditioned physically and mentally by the features of their immediate surroundings: the lay of the land, the rocks, and the soil; the springs, rivers, and lakes; the various forms of animal and plant life; the sunshine and rainfall; all the natural phenomena they experienced directly. Their bodily responses were conditioned and their mental processes were informed by the environmental stimuli they perceived through their senses. They thereby acquired an empirical knowledge that was more holistic than analytic but so precise and so well fitted to their local environment that it enabled them to cope effectively with the various aspects of the wilderness in which they lived, much as wild animals do in their native habitats.

Ever since the development of agriculture in the Neolithic period, the immense majority of human beings have lived in environments they have transformed. As a result, few of us really desire to inhabit the wilderness permanently and even fewer could long survive in it. This is not because we are genetically different from the Stone Age hunter-gatherers, but because humanized environments do not provide the opportunities for the expression of certain human potentialities that are still in us but can be expressed only under conditions similar to those of Paleolithic life.

Recent observations prove that our genetic endowment would enable us, if conditions were right, to acquire the kind of organic, holistic knowledge that Stone Age people derived from their sensory experiences of the natural environments. This happened in the eighteenth and nineteenth centuries to the Europeans who became *coureurs de bois* or otherwise lived out of contact with civilization in various parts of North America. Having to function in environments that had not been domesticated and that were often very demanding, many Europeans developed within a remarkably short time the biological and psychological traits essential for life in the wilderness. These traits were the expression of fundamental human attributes that had remained dormant during centuries of civilized life in Europe.

Even among people today, the really experienced hunters or fishermen perceive with their whole body the layout of the landscapes and waterscapes in which they practice their particular sport. They come to know almost instinctively the habits of the animals that interest them and how these habits are affected by the seasons, the vagaries of the weather, and other aspects of the environment. In *Meditations on Hunting,* José Ortega y Gasset reports how the hunter "instinctively shrinks from being seen" and "perceives all his surroundings from the point of view of the animal." For the hunter, "wind, light, temperature, ground contour, minerals, vegetation, all play a part. They are not simply there . . . as they are for the tourist or the botanist, but rather they function, they act. . . ." These words of Ortega convey how human beings can still learn to function as organic parts of a given environment instead of simply observing it passively, as do most people when looking at scenery. Human beings can even learn about nature from animals. A naturalist who had raised two wolves and was in the habit of taking long walks with them in the wilderness described how the behavior of his animal companions made him perceive aspects of Nature—smells, sounds, and sights—that he had not noticed before. Thus, although we are no longer adapted to life in the wilderness, this is not due to changes in our genetic nature but to social and cultural forces that inhibit the expression of some of our potential.

FROM the beginning of recorded history and even in prehistoric legends, the word *wilderness* has been used to denote barren deserts, deep forests, high mountains, and other inaccessible or harsh environments not suited to human beings, cursed by God, and commonly occupied by foul creatures. Such forms of wilderness evoked a sense of fear for a good biological reason. They are profoundly different from the environmental conditions under which our species acquired its biological and psychological characteristics during the Stone Age.

The word *wilderness* occurs approximately three hundred times in the Bible, and all its meanings are derogatory. In both the Old and New Testament, the word usually refers to parched lands with extremely low rainfall. These deserts were then as now unsuited to human life, and they were regarded as the abodes of devils and demons. After Jesus was baptized in the Jordan River, he was "led up by the Spirit into the wilderness to be tempted by the devil." The holy men of the Old Testament or of the early Christian era moved into the wilderness when they wanted to find a sanctuary from the sinful world of their times. Thus, while some great events of the Judeo-Christian tradition occurred in the desert, this environment was at best suitable for spiritual catharsis.

In Europe, the word *wilderness* applied chiefly to primeval forests, high mountains, and marshes because these parts of the continent were uninhabitable. According to Marjorie Nicolson, people until the eighteenth century regarded mountains as "nature's shames and ills . . . warts, blisters, and imposthumes" upon the otherwise fair face of the Earth. When the Puritans arrived in the New World, the huge forests that covered the Atlantic coast at that time similarly appeared to them as a "heidious and desolate wilderness full of wilde beastes and wilde men." The majority of immigrants who settled in the rest of the American continent during the following two centuries also regarded the primeval forest with fear and contempt.

Ecologists define as wilderness any environment that has not been disturbed by human activities, but in the popular mind, the word still has a deep resonance with a feeling of alienation and insecurity. It is used to denote almost any place, natural or artificial, in which people feel lost or perplexed. In the past, nature in the wild has been usually regarded as alien and cruel, the site of evil and witchcraft. Now, many people in industrialized societies use the word *wilderness* to denote huge anonymous urban agglomerations that appear to them hostile and corrupt.

Humankind has always struggled against environments to which it could not readily adapt; in particular, it has shunned the wilderness or has destroyed much of it all over the world. Contrary to what is often stated, this is just as true of Oriental as of Occidental people.

The admiration of wild landscapes expressed in Oriental arts and literature probably reflects not so much the desire to live in them as the intellectual use of them for religious or poetic inspiration. The ancient Chinese, especially the Taoists, tried to recognize in nature the unity and rhythm that they believed to pervade the universe. In Japan, the followers of Shinto deified mountains, forests, storms, and torrents and thus professed a religious veneration for these natural phenomena. Such cultural attitudes were celebrated in Chinese and Japanese landscape paintings more than a thousand years before they penetrated Western art, but this does not prove that Oriental people really identified with the wilderness. Paintings of Chinese scholars wandering thoughtfully up a lonely mountain path or meditating in a hut under the rain suggest an intellectual mood rather than life in the wilderness. The Chinese master Kuo Hsi wrote in the eleventh century that the purpose of landscape painting was to use art for making available the qualities of haze, mist, and the haunting spirits of the mountains to human beings who had little if any opportunity to experience these delights of nature. Much of the Chinese land had been grossly deforested and eroded thousands of year before, and the Taoist movement may have been generated in part by this degradation of nature and as a protest against the artificialities of Chinese social life.

In the Christian world, also, there has been a continuous succession of holy men, poets, painters, and scholars who did not live in the wilderness but praised it for its beauty and its ability to inspire noble thoughts or actions. St. Francis of Assisi was not alone among medieval Christians in admiring and loving nature. The Swiss naturalist Conrad Gessner wrote in 1541 that "he is an enemy of nature, whosoever has not deemed lofty mountains to be most worthy of great contemplation." After Jean Jacques Rousseau the many romantic writers, painters, and naturalists of Europe became more than a match for the Chinese poets and scholars of the Sung period depicted against a backdrop of mountains and torrents. But like the Chinese, they wrote of the wilderness in the comfort of their civilized homes, as intellectuals who preached rather than practiced the nature religion.

In Europe the shift from fear to admiration of the wilderness gained momentum in the eighteenth century. The shift was not brought about by a biological change in human nature but was the consequence of a new social and cultural environment. Fear of the wilderness probably began to decrease as soon as dependable roads gave confidence that safe and comfortable quarters could be reached in case of necessity. There were numerous good roads in western Europe by the time Jean Jacques Rousseau roamed through the Alps and Wordsworth through the Lake District. In the New World, access was fairly easy even to the High Sierras when John Muir reached them from San Francisco.

Appreciation of the wilderness began not among country folk who had to make a living in it, but among city dwellers who eventually came to realize that human life had been impoverished by its divorce from nature. People of culture generally wanted to experience the wilderness not for its own sake, but as a form of emotional and intellectual enrichment. In Europe, Pétrarch is the first person credited with having deliberately searched mountain and primeval forest for the sheer pleasure of the experience. His account of his ascent of Mount Ventoux in 1336 is the first known written statement of the beauty of the Alps under the snow, but he reproached himself for letting the beauty of the landscape divert his mind from more important pursuits. By the early Romantic period, however, the wilderness came to be seen not only as the place in which to escape from an artificial and corrupt society but also as a place to experience the mysterious and wondrous qualities of nature. The wilderness experience became a fashionable topic of conversation as well as of literature and painting and thus rapidly changed the attitudes of the general public toward nature.

Until the eighteenth century, for example, the Derbyshire peak region in England was considered wild and unfit for human eyes. In 1681, the poet Charles Cotton described it as "a country so deformed" that it might be regarded as "Nature's pudenda." Travelers in those days were advised to keep their coach blinds drawn while traversing the region so as not to be shocked by its ugliness and wildness. Within a few decades, however, the very same region came to be regarded as so attractive that it inspired lines of extravagant praise by nineteenth-century poets. The Derbyshire hills are now considered rather tame, since they do not exceed 2,000 feet in elevation; later poets shifted their admiration to wilder and more rugged sceneries such as the Lake District, the Alps, and high mountains in general. In less than two centuries, new emotional and intellectual attitudes thus completely changed the relationships between English people and their natural environment.

People who express love for the wilderness do not necessarily practice what they preach. In 1871, Ralph Waldo Emerson refused to camp under primitive conditions when he visited John Muir in the Sierras and elected instead to spend the night in a hotel. When Thoreau delivered the lecture with the famous sentence "In Wildness is the Preservation of the World," he was living in Concord, Massachusetts, a very civilized township where the wilderness had been completely tamed. He loved the out-of-doors, but knew little of the real wilderness. His cabin by Walden Pond was only two miles from Concord; woodchucks were the wildest creatures he encountered on his way from the pond to town, where he often went for dinner. In fact, Thoreau acknowledged some disenchantment when he experienced Nature in a state approaching real wilderness during his travels through Maine.

As pointed out by Aldous Huxley in his essay "Wordsworth and the Tropics," the sceneries that inspired Thoreau and nineteenth-century Romantic poets were very different from the wilderness which has frightened people throughout the ages. "To us who live beneath a temperate sky and in the age of Henry Ford, the worship of Nature comes almost naturally. It is easy to love a feeble and already conquered enemy. . . . There are . . . wild woods and mountains, marshes and heaths, even in England. But they are there only on sufferance, because we have chosen, out of our good pleasure, to leave them their freedom." For us, now, "the corollary of mountain is tunnel, of swamp an embankment; of distance, a railway." In the real wilderness, however, "rivers imply wading, swimming, alligators. Plains mean swamps, forest, fevers."

As did their Oriental counterparts, Christian advocates of the wilderness discovered its beauty while trying to escape from their social environment in search of a better way of life. They valued it as a symbol of anticorruption at least as much as for its own sake; the European pro-wilderness movement gained momentum in the nineteenth century from the reaction against the brutalities of the Industrial Revolution.

Appreciation of the wilderness was later enriched by science. Instead of regarding deserts and marshes as the abode of evil spirits, and mountains as ugly deformities of the Earth's surface, educated people learned to look at these phenomena as expressions of a natural order different from the creations of the human order, but with a beauty of their own. Most people probably still experienced awe in the face of wilderness, but they also had a sense of sublimity at the prodigious creativeness of Nature and a feeling of reverence for the laws—divine or natural—that link humankind to the rest of creation.

INCREASINGLY during recent years, interest in the wilderness and the desire to preserve as much of it as possible have been generated by an understanding of its ecological importance. It has been shown, for example, that the wilderness accounts for some 90 percent of the energy trapped from the sun by photosynthesis and therefore plays a crucial role in the global energy system. The wilderness, furthermore, is the habitat of countless species of animals, plants, and microbes; destroying it consequently decreases the earth's biological diversity. This in turn renders ecosystems less resistant to climatic and other catastrophes, and less able to support the various animal and plant species on which we depend. Undisturbed natural environments, including forests, prairies, wetlands, marshes, and even deserts, are the best insurance we have against the dangers inherent in the instability of the simplified ecosystems created by modern agriculture. From a purely anthropocentric point of view we must save as much wilderness as possible because it constitutes a depository of genetic types from which we can draw to modify and improve our domesticated animals and plants.

Admiration of the wilderness can thus take different

forms. It can lead to direct and prolonged experience of the natural world as in the case of John Muir. For many more people, it derives from a desire to escape from the trials and artificialities of social life, or to find a place where one can engage in a process of self-discovery. Experiencing the wilderness includes not only the love of its spectacles, its sounds, and its smells, but also an intellectual concern for the diversity of its ecological niches.

Above and beyond these considerations, however, are moral ones that also favor wilderness preservation. The statement that Earth is our mother is not a sentimental platitude. Our species has been shaped by the Earth and we feel guilty and somewhat incomplete when we lose contact with the forces of nature and with the rest of the living world. The desire to save forests, wetlands, deserts, or any other natural ecosystem is an expression of deep human values. Concern for the wilderness does not need biological justification any more than does opposing callousness and vandalism. We do not live in the wilderness but we need it for our biological and psychological welfare. The experience of the quality of wildness in the wilderness helps us to recapture some of our own wildness and authenticity. Experiencing wildness in nature contributes to our self-discovery and to the expression of our dormant potentialities.

The human species has now spread to practically all parts of the Earth. In temperate latitudes, although not in tropical or polar regions, we have enslaved much of Nature. And it is probably for this reason that we are beginning to worship the wilderness. After having for so long regarded the primeval forest as the abode of evil spirits, we have come to marvel at its eerie light and to realize that the mood of wonder it evokes cannot be duplicated in a garden, an orchard, or a park. After having been frightened by the ocean, we recognize a sensual and mystic quality in its vastness and in the endless ebb and flow of its waves. Our emotional response to the thunderous silence of deep canyons, to the frozen solitude of high mountains, and to the blinding luminosity of the desert is the expression of aspects of our fundamental being that are still in resonance with cosmic forces. The experience of wilderness, even though indirect and transient, helps us to be aware of the cosmos from which we emerged, and to maintain some measure of harmonious relation to the rest of creation.

CHAPTER THREE

Can the World Be Saved?

LONG ISLAND, in New York State, has been occupied by several Indian tribes of the Delaware stock for thousands of years. Europeans established settlements on it early in the seventeenth century and soon developed agriculture, whaling, and various other industries. Yet despite this long history of human occupation, the island remained unspoiled until late in the nineteenth century, as attested by the book *Fishing in American Waters,* published by Senio C. Scott in 1875: "There is not within any settled portion of the United States another piece of territory where the trout streams are . . . so numerous and productive as they are throughout Long Island. It is scarcely possible to travel a mile in any direction without crossing a trout stream, whether from Coney Island to Southampton on the south side, or from Newton to Greenport on the north side." Like the Ilissus in Athens, all the trout streams of Long Island have now been completely ruined by industrialization and overpopulation.

Environmental degradation of human origin has been going on in many parts of the Earth for thousands of years, but the process has been vastly accelerated by the Industrial

Revolution and by the pursuit of economic growth. Some of the worst environments, in fact, were created during the nineteenth century, as exemplified by Charles Dickens in his description of Coketown. Until our times, however, environmental degradation was limited to only a few areas of the Earth. Now it is spreading to all parts of the planet. There is no need to restate once more the gravity and extent of global degradation, but a few aspects of the problem must be emphasized because they are increasing in importance or are only now being recognized.

Air pollution used to be regarded as a local affair, especially manifested by different types of smogs associated with a few large cities and heavy industries. Smogs have long been known to rot people's lungs, to kill the pines of Rome, to drive song birds away from New York City, to erode Cleopatra's Needle in Central Park, the Parthenon in Athens, statues and buildings in all modern cities. But only recently have we realized that air pollutants are carried by winds, in some cases, over the entire globe.

Wastes originating in industrialized areas are threatening some forms of ocean life and probably causing a reduction of photosynthesis in ocean systems. The oxidation of sulfur and nitrogen in power plants and internal combustion engines produces acids that are carried by air currents far from their points of origin and reach the Earth's surface and bodies of water in the form of acidic rains that damage vegetation, alter aquatic life, and cause leaching of certain soil constituents. It has been calculated that if the present concentrations of acids in the rain over New England were to be maintained for ten years, the productivity of forestry and agriculture there would decrease by 10 percent; such a decrease in photosynthesis for that region alone would constitute a loss corresponding to the energy output of fifteen 1,000-megawatt power plants. Similar or worse damage is being caused in Scandinavia by acidic rains originating in Great Britain and Germany.

The reckless use of energy by industrial nations has probably begun to alter the global climate by excess heat production, the accumulation of dust particles, and the increase in atmospheric carbon dioxide. Even so-called "natural" sources of energy—those from water, sun, wind, or the depths of the Earth—are not as safe as commonly believed. Any form of energy used on a large scale can disturb the natural patterns of energy flow through global systems. Industrial plants on both sides of the North Atlantic may soon discharge enough heat into the Gulf Stream to affect the subpolar marginal ice-covered regions. The melting of the polar ice cap by indirect human intervention is thus considered a real possibility. Although it is not possible to predict in detail the climatic changes that would result from continued growth in energy consumption, most experts believe that *global* disturbances can be expected around the year 2000 and significant *regional* disturbances much sooner.

Water pollutants also have a wider distribution than used to be the case. An enormous variety of synthetic chemicals biologically active even in very dilute solutions are now found in most bodies of water throughout the industrialized world and even beyond. Several of these chemicals have been found to be carcinogenic, and many of them have deleterious effects on aquatic life. Dr. Robert Cushman Murphy, a celebrated scientist at the New York Museum of Natural History, recently expressed his distress over the condition of the world's oceans:

> When I went . . . across the Sargasso, from the West Indies to Africa, 58 years ago, lowering my dory almost daily, it was then the purest and most pellucid water in the world, actually more transparent, as determined instrumentally, than any spring, or pool of melted snow, or mountain tarn, and completely devoid of continental dust. But the ocean today has more to degrade it than petroleum. Into it from all inhabited shores go pesticides, herbicides, defoliants, fertilizers, detergent residues, radioactive poisons, and salts from irrigation, not to mention ordinary sewage.

When Thor Heyerdahl recently crossed the North Atlantic in his papyrus craft, he found the Sargasso Sea so polluted that his crew was reluctant to wash in it. Masses and globs of asphaltlike oil extended over a continuous stretch of 1,400 miles.

The pollution of the Sargasso Sea, the North American Great Lakes, and Lake Baikal is more than matched by that of the Mediterranean. Thirty deaths from cholera contracted from shellfish around Naples in 1973, 300,000 cases of salmonellosis recorded per year around the Mediterranean shores, heavy bacterial contamination of sea water around Marseilles are but a few of the manifestations of the Mediterranean disaster. Others, perhaps more threatening in the long run, are oil slicks, detergents, organic chemicals discharged by industrial plants, and high mercury levels in fish that portend a Mediterranean Minamata. Finally, there is the visual pollution created by the arrogant wall of concrete apartment houses along the shores, where the eye could formerly enjoy the enchanting spectacle created by the blue water, the coastal forests, and the windmills of the hill towns.

Over many parts of the Earth, enormous areas of arable land are being lost every year by desertification and erosion. Some deserts are of course natural phenomena resulting from patterns of atmospheric circulation over which we have no control. But desertification, the consequence of human pressures on dry and fragile ecosystems, is another matter. The lands that have thus been transformed into deserts all lie along the edges of the great dry regions in Asia, Africa, and the Americas where the average rainfall is scant and erratic, from 100 to 400 millimeters a year, during a wet season limited to two or three months. In the past, people living in these areas had developed practices of land use and nomadism that permitted the vegetation to maintain itself despite climatic constraints, but the population has increased and the ways of life have changed. The inhabitants of semiarid regions want to enlarge their herds to the maximum since their wealth is measured in the number of head of livestock they own. Increase in population and increase in livestock trigger a vicious circle. The herds first denude the nearest pastures, then use grazing land farther and farther away. The number of animals coming to drink at each watering point also increases; as a result the soil is ruined over a wide, roughly circular area around the well. The area of cultivated land spreads farther afield because there are more and more mouths to feed, but at the same time the yield from the exhausted soil decreases rapidly. This in turn forces farmers to put other plots of land under cultivation. Trees and shrubs disappear, eaten by goats and camels or burnt in domestic hearths. The barren soil is swept away by the wind and the underlying sand is freed to invade nearby land that might still be cultivable. The result is a spectacular advance of the desert. In seventeen years, the Sahara has advanced fifty to sixty miles southward in the Sudan. The Atacama Desert of South America is advancing by one or two miles a year along a front of fifty to 100 miles. The Thar Desert of India has gained half a mile a year for the past half century.

In the famous Serengeti Park of Tanzania, stretching in the southeast to the Olduvai Gorge, rainfall is uncertain, the land poor, the grass cover feeble. Yet the area can support large populations of wild animals because each species has evolved to find its own ecological niche and the different species have complementary food preferences. However, this subtle ecological equilibrium is readily disturbed by human intervention. The Masai people, who keep large herds of cattle, have already overgrazed and laid waste much of the 23,000 square miles of Tanzania they control. They create desertic conditions wherever they move into the Serengeti and bring about the destruction of the wilderness and its wildlife.

Although much of the Middle East is now an arid desert, the region was rich in trees and animals when Moses led the children of Israel through the Sinai wilderness. The mountains above the Promised Land were then cloaked with dense forests; Asiatic lions stalked abundant wildlife in the more open areas. Today one can go for days through that country and never see a living thing. The remote Kammouha Forest in Lebanon, 6,000 feet high in the northern mountains, provides evidence that human activities, not a change in climate, are responsible for desertification. Adjacent to the desert and in the same climatic zone is a handsome mixed forest that still exists because it was protected from exploitation. It is a small Eden, surrounded by the barren stone skeleton of the former Greater Eden, the Promised Land into which Moses once led his people.

Kashmir, the ancient Mogul's "paradise on earth," provides a contemporary example of a desert in the process of creation as a result of unwise land use. Kashmir is a lovely mountain land along India's northernmost border where magnificent wild areas had been set aside as a wilderness park during the period of British rule and with the cooperation of the local maharajah. When independence came, the people used the park as a source of natural resources. The region immediately underwent heavy forest cutting, overgrazing, and large-scale poaching that resulted in the virtual extermination of the Kashmir stag. In less than ten years much of the land changed from a dense conifer forest to a state approaching high-altitude deserts; the vegetation was pulled apart and overgrazed, the result being the loss of the topsoil.

Paradoxically, the tropical rainforest, which is the apparent opposite of the desert, is also undergoing a process of desertification. Of all the world's ecosystems, the tropical rainforest contains the largest biomass and has the greatest variety of animal and plant species, hence its crucial importance for global ecology. But it is rapidly being exploited in all continents where it exists, with consequences that will certainly soon be disastrous unless new techniques of sylviculture are developed to suit tropical conditions.

The tropical rainforest is probably the most ancient ecosystem of our planet, but it is extremely fragile despite its massive accumulation of organic matter. Its soil has been sheltered from solar radiation for millennia and has thus provided an immense diversity of niches for countless forms of life that directly and indirectly contribute to the nutrition of trees and other vegetation. As soon as the trees are cut down and the soil exposed to the sun, humus begins to decompose and is soon destroyed. The soil, becoming dry and hard once exposed to the scorching sun, looks like baked clay and is unsuited to any kind of vegetation or to other forms of life. In the temperate regions, the forest recovers rapidly after being cut down, thanks to moderate temperature and to regular rainfall, but it does not reestablish itself under tropical conditions.

The tropical rainforest plays a vital role in the ecological balance of the globe by trapping enormous amounts of solar energy and consuming corresponding amounts of atmospheric carbon dioxide. Many of its living species will be lost forever when their natural habitats disappear. As deforestation is proceeding at an accelerated rate in all tropical regions, planet Earth is likely soon to be threatened by physical disturbances and biological impoverishment. Although this statement could be illustrated with examples from any continent, it suffices to mention the case of the Brazilian state of Espirito Santo, a once lush area of tropical forest north of Rio de Janeiro, which is on its way to becoming a desert.

A few decades ago, the forest was cut down to provide fuel for Brazil's burgeoning steel industry. The cleared land was used for pastures and coffee plantations for a few years but rapidly underwent a process of desertification. In a period of twenty years, at least 450 species of plants and more than 200 species of birds disappeared from the region because their habitats were destroyed—with the result that pests and parasites now multiply uninhibited and attack any grass, leaves, and fruit that farmers attempt to cultivate. Human beings are both causes and victims of this ecological collapse. Now that the trees are gone, skin cancers are common among light-skinned people who live in the state of Espirito Santo, especially children, as a result of constant exposure to the sun.

Poverty and destructive land use are a recurring pattern in areas undergoing desertification. Deserts will therefore continue to advance into regions of low rainfall unless social factors are rapidly ameliorated. Unfortunately, the people who live in regions where the tropical forest is being destroyed and the desert is advancing do not perceive the danger of the situation to their own future, let alone to that of the globe. Few of them in fact recognize that overgrazing or the cutting down of trees is responsible for the destruction of their land.

INDUSTRIALIZATION and urbanization do more than destroy the wilderness; they also spoil the quality of humanized landscapes. The creation of agricultural lands in the past has usually required profound transformations of primeval ecosystems, but it has usually respected some of their most interesting natural features and has thus often created enchanting landscapes from them. To a large extent, the charm of the countryside has resulted from the ancient management of nature for agricultural purposes. In countless places, however, the aesthetic and other sensual qualities of the environment are now being spoiled not only by urban and industrial development, but also by changes in agricultural practices. A few examples illustrate the extent to which such environmental degradation of humanized landscapes is taking place in the United States and Europe.

The New England countryside owes much of its charm to the harmonious combination of woodland, meadows, cultivated fields, and human settlements, but deplorable changes have recently been taking place. During the past twenty-five years, Vermont has lost 2 million acres of dairy farmland, amounting to 30 percent of the state's area. The farmland has reverted to scrub forest or has been used for recreation developments, shopping centers, and other commercial enterprises—all changes that decrease the aesthetic quality of the landscape. In New Hampshire, the Great Monadnock has long been one of the most admired mountains of the state, almost the equivalent of Paul Cézanne's Mont Sainte-Victoire. It has been painted by countless artists and celebrated by Emerson, Thoreau, Channing, Mark Twain, Kipling, and many other writers in poems, essays, and letters. The mountain is still there, of course, but the countryside is now very different from what it was during the latter part of the nineteenth century. The land was then mostly open and provided wide horizons that are now being lost as farms have been abandoned and trees have reinvaded the whole area. Similar changes in the characteristics of agricultural land are going on in many other parts of the United States, for example in Oxford, Mississippi, where places described by William Faulkner only twenty-five years ago can no longer be recognized: ". . . almost all of the color of his time is gone. Forests dominate the landscape that as recently as 1950 was pasture and cropland."

Similar losses of environmental quality are occurring in Europe. The familiar patchwork of small fields and thick hedges that has dominated the scenery of East Anglia for more than two hundred years is disappearing, as is also the bocage type of country in continental Europe. Yet the hedges provided habitats for song birds and an immense variety of small animals. They were once among the most valued amenities of the European landscape. But they are incompatible with the use of large-scale agricultural equipment and must therefore be sacrificed at the altar of economic efficiency.

Various kinds of wilderness are being destroyed or spoiled all over the world. Laws may prevent exploitation or permanent occupation of wilderness areas, as in the case of national parks, but they cannot protect them against the damaging effects resulting from the mere presence of innumerable tourists. The phenomenal increase in public curiosity about certain wilderness areas makes it increasingly difficult to experience their qualities. The wilderness is thus another one of the worlds we are losing, in this case losing it as experience even though we preserve it physically.

WHEREAS history is replete with environmental disasters, the increase in world population and the destructive powers of certain technologies during the past few decades has magnified the dangers that threaten humankind and the Earth. This increased magnitude amounts to a qualitative change and raises the question, "Can the world be saved?" The implications of this question, however, are not at all clear. For example, nuclear warfare on a global scale would kill or seriously harm immense numbers of people, animals, and plants, and would probably put an end to our present civilization. But whatever its destructiveness, many microbes, plants, animals, and even people would survive. Life would proceed, although under more difficult circumstances. Our civilization might be destroyed but the world would be saved.

In the minds of most people, the question, "Can the world be saved?" refers less to the Earth and its living forms than to our ways of life and to the future of our civilization. We find it extremely difficult, if not impossible, to take any view of the world's problems that is not anthropocentric. It is from a human point of view that I have listed in Appendix I the dangers that, in my opinion, constitute the greatest threats to the Earth and to life on it. Appendix II shows that practically all technological innovations have eventually brought about some form of environmental and social degradation.

It would be easy to elaborate on the dangers listed in Appendices I and II. Much could be added to the list of values we are losing and to the difficulties that are constantly emerging in every part of the world. Yet I have faith in the future because I believe that our societies are learning to anticipate the dangers they will face and to deal with them preventively before irreversible damage is done. Furthermore, I am inclined to agree with Confucius that lighting a candle is better than cursing the darkness. I shall therefore now examine how we can help natural and humanized environments to recover from their present state of degradation and how we can bring out environmental values that would remain unexpressed in the state of wilderness.

CHAPTER FOUR

The Resilience of Nature

Mechanisms of Ecological Recovery

WHEN the United Nations Conference on the Human Environment convened in Stockholm in 1972, its main themes were expressed in phrases such as "pollution is making the Earth unsuited to life"; "rivers and lakes are dying"; "deserts are on the march"; "natural resources are being depleted." This pessimistic view of the world's condition reflected an awareness on the part of the general public that human activities are now causing extensive environmental damage all over the Earth. Deforestation, erosion, and salination are lowering agricultural productivity in many places. Chemical pollutants are spoiling the quality of air, water, and other aspects of the environment, with deleterious effects on all forms of life. Human activities decrease the wealth of the Earth by wasting its natural resources and bringing about the extinction of many living species. Aldo Leopold's words about prairie vegetation could apply just as well to many other ecosystems now threatened by human activities: "No living man will see again the long-grass prairie, where a sea of prairie flowers lapped at the stirrups of the pioneer. We shall do well to find a forty here and there on which the prairie plants can be kept alive as species. There were a hundred such plants, many of exceptional beauty. Most of them are quite unknown to those who have inherited their domain." The fact that lovely wildflowers that used to be common in our woods are becoming extremely rare—the fringed gentian and the trailing arbutus, for example—testifies to the impoverishment of the native flora in many places during the past few decades. Concern for endangered species focuses chiefly on animals and plants that are well known because of some obvious attribute, but the gravest losses may be of species that are essentially unknown yet play an important role in the ecology of tropical or other fragile ecosystems.

Many people believe that much of the damage done to the Earth is so profound that it is now irreversible. Fortunately, this pessimism is probably unjustified because ecosystems have enormous powers of recovery from traumatic damage.

Ecosystems possess several mechanisms for self-healing. Some of these are analogous to the homeostatic mechanisms of animal life; they enable ecosystems to overcome the effects of outside disturbances simply by reestablishing progressively the original state of ecological equilibrium. More frequently, however, ecosystems undergo adaptive changes of a creative nature that transcend the mere correction of damage; the ultimate result is then the activation of certain potentialities of the ecosystem that had not been expressed before the disturbance.

Numerous examples of such environmental recovery occur either through homeostatic response or adaptive changes of a creative nature. I shall describe a few of them, selected because they illustrate environmental problems in different climatic zones and different technical approaches to environmental improvement.

Homeostatic Recovery of Natural Ecosystems

ANYONE who has established a home on abandoned farmland in the temperate zone is painfully aware of Nature's ability to restore the forest vegetation that existed before the advent of agriculture. This has been my own experience in the Hudson Highlands forty miles north of New York City, where endless struggle is required to prevent the return of the type of forest that once covered that part of the world.

A recent bulletin from the University of Rhode Island Agricultural Experimental Station provides a typical illustration of the restorative ability of Nature in the temperate zone. Two centuries ago, 70 percent of the land in Rhode Island had been cleared of the deciduous forest that once covered it almost completely. The primeval forest had been transformed into agricultural land by the original white settlers. During the late nineteenth century, however, the less productive farms were abandoned and trees returned so rapidly that less than 30 percent of the state remains cleared today. Nature provided the mechanisms for a spontaneous step-by-step restoration of the original ecosystem. Similarly, forest is reoccupying abandoned farmlands in many other areas of the eastern United States. For example, although Massachusetts is one of our most heavily populated states it has now become one of the most heavily forested. The return of the trees is not peculiar to the Atlantic coast. In Michigan, the Porcupine Mountain forest, which had been badly damaged by mining during the nineteenth century, has now recovered so well that it is called the Porcupine Wilderness State Park.

Deforestation began in western Europe during the Neolithic period and probably reached its peak a century ago, but brush and trees take over as soon as agricultural land is abandoned for economic reasons; the original ecosystem may become reestablished even on land that had been cleared and cultivated for more than a thousand years!

The native forest has returned even to the 50,000 acres of the Verdun region of northeast France, where the French and German armies fought the most destructive and longest battle of World War I. At the end of the battle, practically all the trees had been destroyed; yet the original vegetation is now back, as are birds, rabbits, and deer. It is believed that the forest will have returned to its original state within less than a century after the battle of 1916.

Similar phenomena of restoration can be observed in tropical and subtropical countries. When the Korean War ended in 1953, a demilitarized zone (DMZ) of 2.5 miles width was agreed upon between North and South Korea. The DMZ was then a wasteland pockmarked with bomb craters and shell holes; yet it has now become one of Asia's richest wildlife sanctuaries. Abandoned rice terraces have turned into marshes used by waterfowl; old tank traps are overgrown with weeds and serve as a cover for rabbits; herds of small Asian deer take refuge in the heavy thickets. Korean tigers and lynx now roam in the mountains of the eastern part of Korea. Birds prosper throughout the DMZ because they are almost completely out of reach of guns; pheasants are so plump that they have difficulty getting off the ground; the rare Japanese ibis has recently been spotted; the monogamous Manchurian crane, a white, black, and red bird with a majestic wingspan of eight feet, can be watched performing its elaborate mating ritual, which consists of bows, wing flapping, and leaps in the air.

Although trees in the temperate zone are the most obvious manifestations of ecological recovery, they are not the only ones. Animals also reestablish themselves as soon as they have a chance. Deer multiply to a nuisance level in all areas where land management provides them with an adequate supply of food; wild turkeys have once more been sighted in all counties of New York State; coyotes and even wolves are on the increase in those parts of the northeastern states that are reacquiring some wilderness characteristics.

The reintroduction of beavers in Sweden provides a picturesque illustration of Nature's ability to reestablish the natural order after it has been destroyed by human intervention. In that country, the last of the original population of beavers was shot in 1871, at a time when the species had disappeared from most of its habitats. However, when a few beavers from the surviving Norwegian stock were reintroduced in Sweden between 1922 and 1939, they rapidly multiplied to such an extent that they caused extensive damage to forest and arable land. There are now once more vociferous demands for an open season against beavers and even for their complete eradication.

In the temperate zone, a forest area of some hundred acres containing a suitable mix of species and habitats can maintain itself under conditions of heavy use and can recover even from severe damage. But this is not the case in tropical, desertic, or arctic regions, nor even for areas that have been stripmined in Ohio, in Pennsylvania, and elsewhere in Appalachia. In many parts of the world, ecologic recovery requires extremely long periods of time and is possible only if very large areas are protected from further damage.

Despite their massive grandeur and seemingly stark immutability, the Himalayas, the Andes, and the East African mountains are among the most fragile ecosystems on Earth. Their steep slopes deteriorate rapidly and often irreversibly when erosion follows such overuse as excessive wood cutting, grazing, or cropping. Semidesertic areas and tropical rainforests also are extremely susceptible to environmental insults. All over the Earth, deserts are indeed on the march. Yet even some of the most fragile ecosystems can recover under special circumstances. Recovery can take place when the proper species reach the damaged ecosystem either by accidental transportation or by active migration.

In 1883, Krakatoa island in the Sunda Strait near the Malay Peninsula was partly destroyed by a tremendous volcanic eruption that killed all its forms of life. Experts have estimated that the explosion had the violence of a million hydrogen bombs. The seismic wave it generated reached 135 feet above sea level, destroying seaside villages in Java, Sumatra, and other neighboring islands. Ash and gases rose 50 miles into the sky, blocking out sunlight over a 150-mile radius. Vast quantities of pumice hurtled through the air, defoliating trees and clogging harbors. When the eruption ended, what was left of Krakatoa island was covered by a thick layer of lava and was completely lifeless.

The wind and the sea currents, however, soon brought back some animals and plants, and life once more took hold on the lava. More than thirty species of plants were recognized as early as 1886. By 1920, there were some 300 plant species and 600 animal species including birds, bats, lizards, crocodiles, pythons, and of course rats. Today, less than a century after the great eruption, the plant community on Krakatoa is fairly complex, although it has not yet reached the composition of the climax forest in the rest of the Malay Archipelago.

Many examples of resurgence of life have been observed under other conditions. For example, the lavas in the Snake River valley of southern Idaho emerged at 2,000°F and were therefore sterile; but after cooling, they were invaded by lichens, which were followed by mosses, then by other plants and a few animals. Even Bikini and Eniwetok, pulverized and irradiated by fifty-nine nuclear blasts between 1946 and 1958, are said to be reacquiring a complex biota, despite the destruction of their topsoil.

The most recent illustration of Nature's biological power is the rapid establishment of living things on Surtsey, the new island created by a submarine volcanic eruption on November 14, 1963, off the coast of Iceland. Within less than ten years after its emergence, Surtsey had acquired from the neighboring islands and from Iceland itself a biota that makes it an almost typical member of the Icelandic ecosphere.

The introduction of biota from an exterior source is not always needed for the recovery of a fragile ecosystem. Many plants or their seeds can persist in a dormant state for long periods of time and prosper again as soon as conditions are favorable for their development. The Wadi Rishrash region of the Eastern Desert in Egypt, for example, was shut off to grazing in the 1920s. Within a few years, the vegetation was so dense that it resembled an irrigated oasis; desert animals took refuge in it during the breeding season. With its new biota, the region appears almost out of place in its barren surroundings.

In Greece and the African Sahel, similarly, a diversified vegetation reappears spontaneously when the land is protected against grazing by cattle, goats, and rabbits; even trees grow in areas that had long been almost desertic. In a particular Sahelian ranch of one-quarter million acres, the land changed spontaneously from the state of desert to pasture when a barbed wire fence was installed to prevent uncontrolled grazing and when cattle were allowed to graze in each area of the ranch only once in five years.

Recovery of a damaged ecosystem occurred recently in west Texas near the city of San Angelo, at the confluence of the three Concho rivers. The process began with Rocky Creek, which dried up thirty years ago but now makes an important contribution to the municipal water supply of San Angelo.

At the turn of the century, Rocky Creek was a neverfailing clear stream that ran through a valley of tall grasses, dotted occasionally here and there by mesquite or some other form of brush. Fish and waterfowl were abundant in the stream; deer and smaller game sought refuge from summer heat under the trees of its banks. Throughout the early decades of the twentieth century, however, brush increasingly invaded the floor of the valley and hillsides. Rocky Creek became progressively narrower and shallower and eventually ceased to flow during the drought of the 1930s. Fish, waterfowl, deer, and other game virtually disappeared.

While the drought contributed to the impoverishment of Rocky Creek, most of the damage seems to have resulted from changes in land use. Before the white settlements, herds of buffalo and other grazing animals periodically migrated down from the plateaus. They were so numerous that they left in their wake a hoof-scarred land almost denuded of grass, but the damage was only temporary because they stayed for a relatively short time; they did not return to the same place for a year, or perhaps for several years, thus allowing grass to grow back. The situation changed when white ranchers made a practice of enclosing cattle within barbed wire fences and keeping them in the same area year after year. As a result of overgrazing, the better types of native grass progressively disappeared and were replaced by mesquite and other brush. The deeply growing roots of these plants sapped the underground water that had formerly found outlets into creeks and rivers and permitted the growth of desirable varieties of grasses.

Around the middle of the twentieth century, a few ranchers began to change their grazing practices in order to protect their cattle against the drought. They reduced the number of livestock so that the level of grazing enabled tall grasses to return and they started a program of brush destruction by herbicides. The result was beyond expectation. In 1964, a half-forgotten spring began to flow for the first time in some thirty years. Its flow progressively increased and soon Rocky Creek also came back to life. As more and more brush was eliminated, new seeps and springs began to flow in the valley. Many of them continued to yield clear water even during a hot, dry summer. Rocky Creek has flowed the year around since the late 1960s, all the way to the Middle Concho. Fish, waterfowl, deer, and other game are once more part of the scenery. Other programs of brush control now conducted by the University of Texas in collaboration with ranchers show that, as was done in the Rocky Creek area, it is possible also in other areas to bring back to life numerous seeps and springs that have been dried up for several decades.

While deserts are on the march over much of the Earth, they can be made to retreat. In 1978, the International Pahlavi Environment Prize was awarded to Professor Mohamed El-Kassas, an Egyptian expert on desertification. Following the award, Professor Kassas made the following statement:

> The desertification process can be reversed. It has been in a number of countries. In the 1930s, the United States experienced droughts that were just as damaging as the Sahel's of the 1970s. In the 1950s, the U.S. faced equally serious droughts, but almost no damage was done. Why? Because meanwhile the Americans had adopted correct land use policies, proper environmental management and the right legislation.
>
> The most important thing is to build up each country's indigenous capacities. This is more important than outside help in the form of food and money. Of course, research is useful, but the basic technologies are already known and, again, it is more important to apply existing knowledge than to enlist more experts. The situation is not at all hopeless . . if we use our resources.

Recovery of Waterscapes

THE recovery of lakes and waterways polluted by industrial and household effluents is another manifestation of the restorative ability of Nature. In several parts of the world, damage caused by water pollution has been completely or partially corrected, not by treating the polluted ecosystem, but simply by interrupting further pollution and letting natural forces eliminate the accumulated pollutants. The results achieved for the Thames in London, the Willamette River in Portland, Oregon, Lake Washington in Seattle, and Jamaica Bay in New York City are but a few among the many examples of improvement in water quality achieved by antipollution measures during the past decade.

In the London area, the Thames River has long been extremely polluted, as attested by Michael Faraday's much publicized letter to the *Times* in 1855. The abundance and variety of fish had started to decrease almost two centuries before in the Thames estuary and only eels survived in certain areas by 1855. As of 1976, however, there were eighty-three species of fish in the estuary and even salmon were caught in London for the first time in approximately 150 years. In the Connecticut River, the Atlantic salmon was reintroduced in 1977 after 200 years' absence and seems to be able to persist.

A few details concerning Jamaica Bay in New York City illustrate the improvement that can be achieved even under the least promising conditions.

Jamaica Bay is a large Atlantic bay adjacent to John F. Kennedy Airport. It used to be the site of an important shellfish industry and offered refuge to hundreds of thousands of migrating waterfowl during the spring and fall, but it suffered extensive damage as a result of its proximity to the large population of New York City and later to the airport. Sand was dredged from its bottom; its water was polluted by discharges from more than 1,600 sewers; the marshes on its periphery were filled with garbage that formed artificial islands.

During the past two decades, however, attempts have been made to save the bay. Water pollution control facilities have been established; the dumping of garbage has been discontinued; grasses and shrubs have been planted on the existing islands of garbage. As a result, shellfish, fin fish, and birds are once more abundant. The center of the bay has become a wildlife refuge. The bird population is remarkable not only for its abundance, but also for its diversity and for the presence of a few unusual species. There are large numbers of wading shore birds such as sandpipers, dowitchers, and green herons. Migration time brings wave upon wave of scaup and brandt, mallards and canvasbacks, Canadian geese and teal. The glossy ibis and the Louisiana heron also have come back, as well as the snowy egret, a bird that was almost extinct in the 1920s.

The return of the glossy ibis to Jamaica Bay, of the wild turkey and the peregrine falcon to their old habitats of the eastern United States, of the salmon to the Willamette River and more recently to the Thames illustrates that once disturbing influences have ceased, some of the original ecological order reestablishes itself spontaneously. Similar phenomena of ecological recovery have been observed in many other parts of the world, particularly in North America and Europe. It is probable therefore that environmental degradation can be interrupted in many cases and that the rate of improvement can often be more rapid than is commonly believed. With good management and human commitment, nature often takes over and heals itself.

Evolution of Natural Ecosystems

NATURAL ecosystems are profoundly different from those of earlier geological times, and even from those of more recent times when the Earth's climate had become essentially what it is today. Ecosystems constantly evolve under the influence of physicochemical forces that are poorly understood but are known to be much more powerful than those unleashed by human activities. Long before the human presence, for example, stupendous dust storms repeatedly occurred on Earth, as they have recently been found to occur on the planet Mars. Climatic changes, volcanic eruptions, earthquakes, hurricanes, fires of natural origin, and the activities of animals and plants have also played their part in the formation of natural ecosystems. During the first half of this century, the evolution of ecosystems was commonly regarded as occurring through an orderly succession of plant and animal species following each other in a fairly well-defined order. According to this view, the end result is a climax population that remains fairly stable until affected by some major disturbance. Reality, however, is far more complex and more interesting than suggested by this simplistic picture of systematic succession and climax. Even without human intervention, many random events influence the evolution of ecosystems in ways that are not predictable.

Fire, for example, was considered for a long time to have only destructive effects, but it is now known to be essential for the development of certain plant species. Small fires prevent the accumulation of fuel on the ground and thus minimize the danger of catastrophic wildfires. The small fires release mineral nutrients from organic debris and make them available for further plant growth. The National Park Service has recently adopted a policy of letting fires of natural origin run their course almost unchecked in certain wilderness areas and even starting controlled fires wherever necessary for the maintenance of certain species of trees or for the health of the land.

In the arid cattle lands of southern Arizona, fire prevention programs have resulted in the establishment of shrubs and trees like mesquite and cholla at the expense of grass. Once this new vegetation was established, moreover, it utilized and reduced the moisture supply, thus preventing the return of the grass even when grazing was discontinued.

Fires have contributed profoundly to the evolution of ecosystems in several parts of the world by interfering with the spread of the forest or destroying it where it existed. This is particularly true for the Great Plains of North America, where preagricultural Indians set fires to facilitate the hunt and thereby prevented almost completely the growth of trees and shrubs. The short-grass prairie might have emerged spontaneously in the Far West, since trees have difficulty in maintaining themselves in areas of low and erratic rainfall, but fire was essential for the creation of the tall-grass prairies in the eastern part of the North American grassland.

Another essential element in the development of the prairie ecosystem was its animal population. Until the turn of the century, immense herds of buffalo trampled open spaces in the grass and at the same time richly manured the soil. In the words of a nineteenth-century observer, the buffalo "press down the soil to a depth of 3–4 feet. . . . all the old trees have their roots bare of soil to that depth." The spaces opened in the grass by the buffalo were utilized by smaller animals. Prairie dogs, for example, supported predators such as the black-footed ferret and also turned over enormous quantities of earth by their incessant burrowing activity.

Whatever the factors involved in the evolution of prairie vegetation, the final result was a balanced system of luxurious and tall grasses, numerous species of wildflowers, and a black sod more than ten feet deep in certain places. So many random events contributed to the emergence of the American prairie that we probably could not recreate it today even if all its plants were known. The flora could not be reestablished in its original state without the participation of all the former fauna and other natural forces, including the trampling of the soil by immense herds of buffalo.

THUS human beings have probably never been in real "balance" with their environments except under conditions where population density is extremely low, as in the polar regions or the Australian desert. Where the land has long been continuously occupied, the analogues of natural ecological communities have progressively emerged over many generations by a trial-and-error method usually guided by unconscious human values.

Most of these communities have undergone profound changes in the course of time as a consequence of wars, epidemics, new agricultural practices, and the conscious or accidental introduction of new animal and plant species. The region of England now known as the Downs, for example, has experienced several very different ecological states during historical times.

This region was completely wooded before human occupation, but Neolithic farmers began to cut down the trees almost five thousand years ago, using fire and stone axes, and most of the forest was progressively converted into farmland. During the middle of the fourteenth century, however, large areas of arable land were abandoned as a result of the plague (Black Death) that decimated the human population. If the epidemic had occurred in earlier Saxon times when the land was cultivated by individual farmers, many fields would certainly have returned to the original forested state, but the feudal system created a different situation. Shepherds with flocks of sheep could deal with much more land than plowmen with ox teams. This change in agricultural practice enabled the lords to keep their land in productive use and free of trees, even where labor was in short supply. Sheep grazing, unlike the munching of cattle, crops the grass to a lawnlike texture. It stimulates the growth not only of grass but also of many wildflowers—rock rose, wild thyme, scabious, and the like—that make the Sussex Downs smell, in the words of Rudyard Kipling, "like dawn in paradise." A multitude of insects, especially butterflies, also flourish on these sheep-grazed plants. Even now, trees come back spontaneously whenever sheep and rabbits are removed from the Downs, but in practice these animals are present almost everywhere and prevent the return of the forest. Thus in this case, as in the case of the American prairie, the creation and maintenance of an artificial yet highly desirable ecosystem is dependent on a multiplicity of random factors resulting in the control of tree growth by the animal population.

Other desirable ecosystems, in contrast, have emerged from the proliferation of certain trees. For example, the Adriatic coast, from the Po valley to south of Ravenna, used to be covered with oak and beech. During the fourth and fifth centuries, monks introduced *Pinus pinea* for its nuts and for aesthetic reasons. The pines thrived, escaped, and established themselves on the hills. Thus was created the harmonious mixture of broadleaf trees and pines that Dante describes in *Purgatorio.*

RAINFALL, wind, and drought have of course exerted an enormous influence on the natural evolution of soils, for example, in the "dust bowl" of the southern Great Plains. It had been feared that the agricultural value of the land would be destroyed by the tremendous dust storms of the 1930s, but in fact the dust bowl region has produced bumper crops during the past two decades, in part as a result of wiser agricultural practices, but also because of somewhat greater rainfall. As mentioned earlier, dust storms have occurred repeatedly under natural conditions in the past and they certainly will occur again in the future. Rainfall is one of those random events that has an unpredictable influence.

Other factors in the evolution of natural ecosystems are changes that have occurred in the very chemical composition of their soils, chiefly but not exclusively as a consequence of agricultural use. In many places, particularly in what used to be the American prairie, the soil has lost much of its organic matter since it has been put under cultivation, but there are other places in contrast where agricultural practices have brought about an increase in soil organic matter. European farmers recognized long ago that most of their soils (of the gray-brown podzolic types) have to be manured for satisfactory crop yield. Similarly, early colonists of New England learned from the Indians to put a fish into each hill of corn. Many gray-brown podzolic soils of Europe and perhaps of the United States now seem richer in organic matter and are more fertile than they were in their original forested state as a consequence of fertilization. Agricultural practices, and the general policies of land management, may now play a role as important as that of natural forces in the evolution of ecosystems.

In most parts of the Earth, ecosystems have thus continuously evolved, first through the influence of random natural events, then as a side effect of human activities and increasingly because of deliberate social choices. Humanized environments could be very different from what they are, but this does not mean that they can safely be anything. The next chapter deals with ecosystems created by deliberate human action that proved viable because they were designed to fit prevailing ecological conditions.

CHAPTER FIVE

Humanization of the Earth

The Wooing of the Earth

WHEN the Bengali poet Rabindranath Tagore (1861–1941) first traveled as a student from India to England in 1878, he realized immediately that the visual charm and the agricultural productivity of the European countryside were the result, in his words, of "the perfect union of man and nature, not only through love but also through active communication." Traveling by railroad from Brindisi to Calais, he "watched with keen delight and wonder that continent flowing with richness under the age-long attention of her chivalrous lover, western humanity." For him, the shaping of the European continent by human labor constituted the "heroic love-adventure of the West, the *active wooing* of the earth" [italics mine]. Yet it is unlikely that Tagore fully appreciated the extent to which the countryside he saw from the train had been shaped by more than a hundred generations of peasants out of the forests and marshes that covered most of western Europe before human occupation.

When first seen by Tagore, Europe presented some of history's most successful landscapes created by the interplay between humankind and Earth. There was still some genuine wilderness, but it had been pushed back to mountaintops and other areas not suitable for human settlements, where it could be admired in safety. Most of the land had been settled, but except in a few places, it had not yet been abused by industrial civilization and by formless urbanization. Means of communication provided easy access to practically all parts of the land, integrating them into organic units but without destroying regional individuality. Villages and towns had been built with instinctive grace using local materials that made them appear to be an expression of nature. The quality of blessedness that emerges from long periods of intimate association between human beings and nature was an essential part of my own experience at the beginning of this century. In the Île de France region, where I was born and raised, what is called nature is profoundly different from the original wilderness. It has been progressively created since Neolithic times by the work of the peasantry—in Tagore's words, by the "active wooing of the earth."

The Île de France was almost entirely covered by trees in the Stone Age, and there is no doubt that it would rapidly return to a state of forested wilderness if it were abandoned. Although the region is commonly regarded as having charm and elegance, its qualities are largely the result of human management. The hills have such low profiles that their only remarkable features are the diversified farmlands and the carefully managed woodlands they now support. The streams are small and sluggish but their shores have been polished and are commonly associated with gentle pastoral scenes. The sky is rarely spectacular and indeed often cloudy, but the climate and the soft luminosity are favorable to an immense diversity of vegetation, much of which either has been introduced from other parts of the world or has been transformed by domestication. Villages with venerable churches crown the summits of the hills or nestle in the valleys, making the human presence everywhere a part of the scenery.

Since the clearing of the primeval forest that began in Neolithic times, the Île de France has acquired a humanized quality that transcends its natural endowments. It has repeatedly experienced destructive wars and social upheavals yet has constantly supported a high population density. Its land has remained fertile and has provided a home for many different forms of civilization. Humanizing the Île de France has admittedly resulted in the loss of many values associated with the wilderness. From the human point of view, however, at least according to my taste and that of many other people, the region is now visually more diversified and emotionally richer than it was in its original forested state. It provides a typical example of what I shall discuss later as the symbiosis of humankind and Earth.

I have used the case of the Île de France to introduce the theme of humanization of the Earth only because this is the region of my birth and upbringing, but what I wrote about it applies to many other parts of the world as well. In most places where human beings have settled, they have created out of the wilderness artificial environments that have become so familiar that they are commonly assumed to be natural although they have a cultural origin. Every continent can boast of "cultural environments"—to use an expression translated from the German and recently introduced into the ecological jargon—that have remained fertile and attractive for immense periods of time and have long been the true homes of humanity.

While the wilderness is still being destroyed in several parts of the world, the most extensive destruction occurred many centuries ago. Surprising as it may seem, a large percentage of the Earth's surface in the Old World was transformed by ancient peoples working with primitive tools and a few domestic animals. The process began during the Stone Age when the agricultural lands of the very first civilizations were created from the Mesopotamian wilderness between the Euphrates and the Tigris. The humanization of the Earth has continued ever since; it was essentially completed in most of Europe and Asia during the eighteenth century. Depending upon the places, it involved deforestation, drainage, irrigation, or such spectacular changes in topography as the terracing of slopes in hilly regions and the reclaiming of land from the sea, as in the Netherlands.

Destruction of the wilderness was delayed in the New World as long as population density remained much lower than that of other parts of the Earth, although preagricultural Indians contributed to the process by the forest fires they repeatedly set in the western parts of the continent. The destruction of the American wilderness proceeded rapidly after the arrival of Europeans. When the United States Census Bureau reported the close of the frontier in 1890, immense areas of North America had been deforested, the "breaking" of the prairie was essentially completed, and large irrigation projects were on the way.

The peasantry of the Old World and the settlers of North America thus created out of the wilderness the cultural environments that constitute most of what we now call Nature. As a result, the various regions of the Earth acquired their distinctive characteristics from agriculture and social institutions, as much as from geology, topography, climate, and rainfall. In other words, except in places that have remained in the primeval state of wilderness, the word *nature* implies human as well as physical geography.

In the temperate zone, a typical humanized landscape consists of pastures and arable lands in the low altitudes and on gentle slopes. Forests occupy almost exclusively the higher altitudes and other areas unsuited to agriculture, industry, or human habitation. Most of the bodies of water have been confined within well-tended banks, controlled by dams, rechanneled, or disciplined in other ways. Despite all this human ordering, we forget that these typical sceneries bear little resemblance to what they would be without human management. We have lived in intimate familiarity with them so long that we contemplate them in a mood of casual acceptance and reverie without giving thought to their origin and evolution. We even forget that most villages and cities are on sites first occupied by human settlements centuries or millennia ago and that roads, highways, and railroad tracks commonly follow trails first opened by hunters, pastoralists, and farmers ages ago.

Many of the animals and plants in humanized landscapes differ from those of the original wilderness, either because they belong to species introduced from other parts of the world or because local species have been genetically modified by selection or other biological manipulation. It is through human agency that wheat, corn, rice, barley, potatoes, tomatoes, oranges, grapefruit, and countless other crops are now cultivated far from their place of origin; that eucalyptus trees grow in California, Italy, Greece, or North Africa, often more vigorously than in their native Australia; that African violets adorn homes in most parts of the world, whether communist or capitalist politically. The tulip, now thought to be characteristic of Holland, was actually introduced there from Turkey. Almond trees, fig trees, and olive trees, which call to mind the Mediterranean region, in fact originated in Asia. One of the most enchanting Greek landscapes is a large olive grove in a valley near Delphi, which must be more than three thousand years old since it is mentioned in Homeric writings.

Above and beyond the agricultural production made possible by destruction of the wilderness, the clearing of the forest generated environmental values that have now become part of the human view of nature. It exposed the architectural skeleton of the various geologic strata—for example, the contrasting structures and textures of white chalky formations or of granitic boulders covered with lichens and mosses. Within a given ecological system, partial deforestation created furthermore an environmental diversity that provides nourishment for the senses and for the psyche. A very human harmony exudes from the mosaics of cultivated fields, pastures, and woodlands, as well as from the alternation of sunlit surfaces and shaded areas. The humanization of the Earth has created, between clearing and woodland, a ruffled skirt of brush rich in brilliant fruit and bird life. My childhood in the farming country of the Île de France has left me vivid memories of moods that ranged from the joyfulness of the lark's call rising from the wheat fields to the quiet of adjacent woodlands.

HUMAN intervention has profoundly transformed the surface of the Earth even in the most unlikely places. As mentioned earlier, the tall grass of the prairie that used to cover part of the North American continent emerged as an indirect consequence of forest fires set by preagricultural Indians. The moors of the British Isles, which have inspired so much literature, do not represent the original natural system of the region; they progressively developed after the deforestation that began during Neolithic times and was maintained by the population of sheep and rabbits.

Human intervention has also helped to create many of the most beloved and productive landscapes of the world. In Tuscany and Umbria much of the land was molded by peasants who rounded the hills and shaped the slopes to create an architecture of terraces.

In northwestern Europe, hedgerow and bocage types of farming country were created either by law or for a variety of other purposes: marking ownership, establishing drainage systems, protecting crops from the wind, and so on.

In southern China, the very artificial "water and mountain" landscapes are among the most monumental sceneries of the world and also the most productive of edible animal and plant life.

In the agricultural areas of the island of Kyushu and other agricultural parts of Japan, trees and land seem to be trimmed to human specifications, measured to human scale. Visitors to the Islands of the Rising Sun in the nineteenth century were amazed to find them laid out as an all-embracing park with farms, villages, and temples beautifully interspersed and integrated.

Although the conscious transformation of the wilderness is more recent on the North American continent than in Asia or Europe, it has taken similar directions. The villages of New England, with their greens and open fields cozily nestled in the valleys, could not have come into existence without the clearing of the primeval forest. In Lincoln, Massachusetts, the low wetlands were kept free of trees by periodic controlled flooding, a practice that farmers called "flooding the meadows." Destruction of the wilderness in that region meant disappearance of certain game birds, such as the wild pigeon and the wild turkey, but the creation of open areas created habitats for many song birds: bobolinks and meadowlarks in the fields, orioles and bluebirds in the orchards, warblers darting among the treetops, swallows snatching insects from the air. The abundance and availability of low, tender new growth in the cutover forest allowed a spectacular multiplication of white-tailed deer—to the point of overpopulation.

The Pennsylvania Dutch country, too, was completely forested three centuries ago and has now been transformed into manicured fields. In the Lake Saint John region of Canada, consisting chiefly of sand plains and granitic outcrops, periodic burning keeps out the forest and favors the growth of blueberries, which find a profitable outlet on the Montreal and New York markets. All over the central and western plains and the deserts of North America, the industrialization of agriculture has led to the creation of humanized landscapes of gigantic dimensions that fit the enormous size of the original ecosystems. Over much of the world, farmland has become the most distinctive feature of the scenery. It constitutes the "Nature" that has replaced the wilderness in the minds of both rural and urban people.

Environmental Needs of Human Life

HUMAN settlements now exist at practically all latitudes, but human beings are biologically out of place in most of the natural settlements where they have made their homes. The reason for this lack of environmental fitness is that our species emerged in a subtropical climate where it acquired certain fundamental biological characteristics that it has retained ever since, whatever the natural conditions under which it lives now. In arctic regions as well as in the tropics, *Homo sapiens* remains genetically best adapted to a certain type of semitropical savanna. We could not survive long even in the temperate zone if it were not for our ability to use fire, build shelters, practice agriculture, and manufacture a great variety of artifacts—thus creating humanized habitats out of the wilderness.

The precursors of our species probably dwelt in a tropical arboreal environment either in East Africa or somewhere in Asia. It is probable that an important factor in their further evolution was a change in climate that caused a reduction of forest density in many places approximately 15 million years ago. Where the dry season was short and interrupted by showers, vegetation consisted chiefly of evergreen forests, but the vegetation took a seasonally deciduous character as the length of the dry season increased. The longer the dry season, the longer the period of leaflessness and the greater the mean distance between the trees. A region where the mean distance between trees exceeds canopy diameter is called a savanna. If its climate is extremely dry, a region takes a desertic character with vegetation dominated by spiny shrub except during the brief rainy season.

Vegetation is most abundant in densely forested areas, but the foliage and fruit are then generally produced high above the ground and therefore cannot be reached by a land-based animal. Vegetation is more readily accessible in desertic areas but is scarce and sporadic. In contrast, photosynthesis in the savanna is chiefly performed by grasses and forbs that provide abundant foodstuff for grazing animals and indirectly for humans who feed on them. Furthermore, many savanna plants store food underground in stems, roots, and tubers that are readily used by both animals and human beings. The savanna type of country therefore provided early peoples with diverse and abundant food resources from plants and animals. It also had streams and lakes where drinking water and fish could be readily obtained. For these reasons, early human settlements were commonly located on the shores of bodies of water.

Life in the savanna had the additional advantage of making early humans less vulnerable to attack by large predators, which cannot be seen at great distance in the forest but are more readily visible in open country. Although many animals were better endowed than people with regard to smell, hearing, speed, and size, the erect posture and good vision of humans made it possible for them to locate predators or prey at a great distance and to devise appropriate methods of escape or attack. The savanna thus constituted an environment favorable for security and for the development of hunting skills.

The experience of several millions of years of life in an open environment where good visibility was essential for survival and for hunting left a lasting stamp on human nature. During the Ice Age, for example, Neanderthal and Cro-Magnon people settled in valleys rich in game and fish. They took shelter in caves or in dwellings built from branches and animal hides, which were located in places from which animal life could be readily observed. Hundreds of such Old Stone Age settlements have been recognized along the valleys of the Dordogne and the Vézère in France. From the huge Cro-Magnon cave at Les Eyzies, the eye can survey a vast panorama of earth, river, and sky. This diversified environment probably created in *Homo sapiens* two different but complementary types of visual conditioning: on the one hand, the need for open vistas leading the eye to the horizon; on the other hand, the need for a place to take refuge, for example, a cave or a densely wooded area, giving protection in case of danger. Children may be displaying this early conditioning when they play hide and seek.

The genetic constitution given to us by the physical characteristics of our evolutionary cradle has not changed significantly during the past fifty thousand years. For this reason, our environmental needs still reflect Stone Age conditions of life. They explain many of our present behavioral patterns, for example:

- the almost universal and subconscious fear of the forested wilderness, an environment where good vision is of little use against danger;
- certain features of design that are common to all schools of landscape architecture;
- the preference of all human beings for the same narrow range of environmental temperature; even Eskimos keep themselves at a semitropical temperature in their well-insulated dwellings and garments;
- the biochemical similarity of nutritional requirements in all human groups; whether carnivorous or vegetarians, all

humans require the same chemical constituents in their diets;

◂§ the fact that practically all the plants we cultivate belong to sun-loving species (as do the plants growing in savanna kinds of country) and cannot therefore grow in the shade of a dense forest.

In fact, human beings rarely make their homes in completely forested areas if they can avoid it. Almost everywhere on Earth since Neolithic times, they have cut down trees to create farmland and to establish their settlements. When compelled to live in a densely forested area, they usually settle in a natural or artificial clearing, preferably close to a river or a lake. To the extent that they can, human beings try to imitate the fundamental features of the savanna environment that was the cradle of our species and to which we are still biologically adapted. Because of its unchangeable characteristics, human life implies the humanization of the Earth.

Homo sapiens *and Nature*

THE preceding discussion should make clear why this book does not deal with environmental problems considered from the orthodox ecological point of view. Its subject is the interplay among us human beings and those aspects of our environment that we perceive and alter to suit our needs and fantasies—an interplay that profoundly alters in turn our own personalities and our ways of dealing with the Earth. This anthropocentric attitude has been repeatedly criticized, but I consider it valid nevertheless and in any case unavoidable. When the critics of anthropocentrism scornfully state that certain environmental practices will make the Earth suitable only for rats, roaches, and ragweed, they express a human judgment that vermin and weeds do not deserve to inherit the Earth—even though most of these species are endowed with remarkable biological attributes. In evaluating ecological situations, we cannot avoid making judgments of values and we naturally give preference to human values.

I shall return to the topic of anthropocentrism in the following chapter, where I express my belief that we human beings can improve on Nature by manipulating it with respect, imagination, and intelligence. At this point, I shall limit myself to a discussion of the argument that modern humankind has become destructive of environmental values because it has lost the quality of relationships to the Earth that animals possess and that guided the environmental behavior of ancient peoples.

A frequently heard criticism of the anthropocentric attitude is that it takes human beings out of the natural order and places them above it, whereas the human species is but one among countless others and should not be given special consideration. Admittedly, we are part of nature like all other living organisms in the sense that we cannot be completely separated from our environment. But it is also true that each living species and each particular organism within a species constitutes an entity distinguishable from the rest of Nature. This is particularly true for the human species because its evolution has been almost entirely cultural—not biological—since the Stone Age. As an animal, *Homo sapiens* is biologically very similar to the great apes but differs profoundly from them in sociocultural characteristics. Paraphrasing Paul of Tarsus, it can be said that humankind is *in* Nature but no longer quite *of* Nature.

Most of us most of the time would have to change our perception of reality not to separate human life from the rest of Nature. We are qualitatively different from other living things not in our anatomical structures and physiological functions, but in our endowment with self-awareness and purpose, and we are culturally shaped by the social structures in which we live.

Another criticism of the anthropocentric attitude is that it disregards the ecological attitudes of ancient times when —or so it is said—people behaved as organic parts of nature and fitted in the natural course of things just as animals do. This argument is almost meaningless even for animals.

Living things must of course function as parts of Nature, but they are never passively molded by their environment. Each of them responds adaptively to the challenges of the local environment in its own particular way. Species that have evolved under the same environmental conditions nevertheless become different from each other because they adapt by different mechanisms, some of which involve profound environmental disturbances. Beavers, for example, create environments suitable to them by gnawing down certain kinds of trees to build dams and flood large areas of land. Many animals adapt to cold during the winter by hibernating in holes that they dig in the ground. The so-called prairie dogs (in reality a kind of large squirrel) used to build enormous underground settlements in several parts of the American West; it has been estimated that some 300 million of these animals were living in burrows within 25,000 square miles in one particular area of Texas at the end of the last century—a veritable megalopolis. On the other hand, bison were prone to break up the mounds of prairie dog settlements with their horns and hoofs, wallowing in the rubble. In Africa, certain species of termite erect huge mounds of earth to house the countless members of their colonies. These few examples suffice to illustrate that in their normal lives, many species of animals profoundly disturb the physical environment.

The lives of predators naturally depend upon their ability to kill the right kind of prey. It used to be believed that lions, tigers, wolves, and other large predators killed only the animals they needed for sustenance, and usually the weak ones at that. However, it is now known that they often kill many more than they need, seemingly in a senseless way, as if for the mere pleasure of killing. Although large apes are primarily herbivorous, they also kill other animals, even of their own species. It takes a Panglossian faith that everything is for the best in the best of all possible worlds to assume that the evolutionary mechanisms of nature have generated in animals the behavioral patterns best suited to the welfare of the planet and the living species it harbors.

There is no evidence either that early humans always lived in ecological harmony with Nature out of respect for it. Even in the Stone Age, humans probably considered themselves somewhat apart from Nature and superior to animals. The oldest portrait of a man so far discovered was drawn some 15,000 years ago on one of the walls of the Trois Frères cave in the south of France. He is commonly referred to as the Sorcerer because his posture and weird accoutrement suggest that he is engaged in some sort of magic. The picture is located in the darkest recess of the cave on a ledge twelve feet above the ground, a location from which the man could have seen the animals depicted on the other walls below him. The placement of the Sorcerer so high on the ledge seems to symbolize that even at this early stage of social evolution, human beings differentiated themselves from the rest of Nature and observed it either to satisfy their curiosity or in a spirit of domination. Climbing a tree or sitting on a rock or mountain to watch the world above and below have been very human things to do throughout the ages.

Humans intervened into Nature, violently and often destructively, even before the beginning of agriculture and herding. The fires set by preagricultural peoples all over the Earth prevented the forest from maintaining itself in many areas where soil conditions and rainfall favored the growth of trees. Later it became the practice, and still is among societies that have not been influenced by industrial civilization, to kill trees in order to cultivate crops. As far back as the Stone Age, deforestation caused by humans had already changed the face of the Earth. Like animal predators, Stone Age people also killed more game than they needed for food, as illustrated by the enormous numbers of horse skeletons accumulated at the base of the Solutré cliffs in central France. It has been claimed indeed that Stone Age people were responsible at least in part for the extinction of several large animal species at the end of the Pleistocene period. In South Island, New Zealand, a primitive Maori population that never exceeded twenty-five thousand people managed, in thirty generations, to bring about the extinction of the eagle, the rail, the goose, the swan, and other birds including the moa, a bird so valuable as a source of food that its name means "great treasure" or "principal resource" in the Maori language.

Some of the worst ecological disasters of human origin

occurred in the earliest civilizations. The inhabitants of Mesopotamia ruined their environment by pushing agricultural productivity beyond what the local circumstances could bear, the result being the erosion of fertile soil into the rivers and the silting up of harbors. The mountains of Lebanon that were famous for their magnificent forests at the time of the Phoenician civilization are now almost barren; in fact, they were robbed of their best trees by the Egyptians and Romans.

Deforestation and erosion began long ago in South America. It forced the Inca empire in Peru to move into the higher valleys of the Andes and to build there the now much admired bench terraces in an effort to save the soil. Mexico also had a serious problem of erosion at the same period. Because of deforestation and erosion, Peru and Mexico were fighting a losing battle with Nature and would eventually have become impoverished even if the Spaniards had not come to conquer them.

It is widely stated that Oriental people have always exhibited more reverence for Nature than Westerners do, but there is little evidence for this in their attitude toward the wilderness. The indiscriminate cutting of trees had been so extensive in many areas of China that in the early days of Chinese history officials repeatedly warned aganst the dire consequences of deforestation in the mountains. Visitors to China in the nineteenth century commented on the absence of trees in the north and the damage done by soil erosion to the loess-covered plateaus, which used to be wooded.

The forests of China were destroyed for a multiplicity of reasons: to create farmland, to make charcoal for metallurgic and other industries (the sky of certain Chinese cities was black with the smoke of pottery kilns), to provide timber for urban construction, to deprive dangerous animals and bandits of hiding places, and on a much larger scale than can be readily imagined, simply to cremate the dead and manufacture the India ink needed by scholars.

Buddhist monks in particular seem to have been responsible for a large percentage of the deforestation in China, Japan, and Tibet. They used enormous amounts of timber for the construction and constant rebuilding of their huge halls and temples. They cremated their dead. The land surrounding the lamaseries was almost made bare by overgrazing by livestock, which were kept even though the monks were supposed to be vegetarians. As early as the fourth century B.C., the Taoist Chuang Tsu looked back nostalgically to the golden age of the past.

Even if it is true that humans became more and more prone to alter the surface of the Earth as time went on, this was not because they ceased to regard themselves as part of Nature, but simply because the world population constantly increased and the means of destruction became more powerful. The change in attitude of the plains Indians illustrates the fallacy of attributing the low level of environmental damage done by ancient peoples simply to their identification with nature.

Until the seventeenth century, the plains Indians derived most of their material needs from bison without significantly affecting the bison population, because they killed only the small number of animals they needed. After the seventeenth century, however, two changes made the hunt much easier. First, the Indians learned to ride mustangs—horses that had multiplied in the wild over much of the continent after being abandoned by the Spaniards—and somewhat later they acquired firearms. With horses and guns, the Indians galloped from all directions into the plains, killing enormous numbers of bison. As the herds dwindled, strife arose over hunting territories and the plains Indians became more warlike. The wholesale destruction of beavers by the Indians for the sake of the European fur trade has been given a complex anthropological explanation, but it was made possible in fact by the availability of firearms.

A somewhat similar situation is now developing in East Africa. As is well known, the game lands of East Africa, particularly the national parks of Uganda, Kenya, and Tanzania, constitute a unique area—seemingly a terrestrial paradise—because of the wealth of fauna and pleasant climatic and other environmental conditions. It is almost certain, however, that this paradiselike state will not last long. Elephant, lion, leopard, rhinoceros, hippopotamus, giraffe, buffalo, gorilla, chimpanzee, antelope, and other large animals are threatened with extinction even in the national parks because most Africans prefer agricultural lands to the wilderness and its wonderful animals, even when they are told that tourism would provide them with a better source of income than agriculture.

Humankind and Nature

As members of the animal kingdom our behavior toward Nature is thus not very different from that of animals. Like the beavers, we ruin land to satisfy our immediate needs, like the big cats we kill animals just for the fun of it, like the lemmings and other species that periodically undergo population crashes, we often use our resources recklessly as if we had no concern for others or for the future. But this is only the animal aspect of human nature; humankind is something more than biologic *Homo sapiens.* Cultural evolution has progressively led us to recognize that the humanization of the planet can be lastingly successful only if fundamental ecological laws are respected.

Tagore's use of the phrase "wooing of the earth" suggests that the relationship between humankind and Nature should be one of respect and love rather than domination. Among people the outcome of wooing can be rich, satisfying, and lastingly successful only if both partners are modified by their association so as to become better adapted to each other. Furthermore, the outcome is more interesting when both partners retain elements of their individuality—of their own wildness.

The same criteria are applicable for a successful association between humankind and our planet. It is because these criteria are rarely fulfilled that few human settlements have long remained in a tolerable condition of ecological health and prosperity. Fortunate are the people, among whom I am one, who have spent their formative years in places where human beings and the planet have long been in close association, experiencing difficulties of course as is the case for all associations, but also continuing to create new values. Having lived in several old countries that still provide vital, pleasant human settlements helps me to have confidence in the future of both humankind and the planet.

In human settlements that have been lastingly successful both the people and the environment have continuously undergone reciprocal adaptive changes. Examples in many different parts of the world show that this process of change is resulting in the progressive humanization of the planet.

Conversely, there has also been going on what could be called a planetization of humankind, which began when Stone Age people changed from hunting-gathering to agriculture.

Whereas hunter-gatherers did not need to look further ahead than a single year, Neolithic farmers had to practice long-range planning to select strains of domestic animals and plants. Agriculture, moreover, eventually implied a certain degree of permanence of a culture. This increased the identification of people with a particular place, a process that may have resulted in a greater diversity among civilizations. Agriculture also made farmers aware that certain of their practices changed the character of the land, for good or for bad, and gave them an empirical knowledge that eventually led to a greater understanding of ecology. It seems very likely therefore that agriculture resulted in a broadening of the human mentality, but perhaps at the cost of some loss in the direct perception of nature—the ripening of berries, the snapping of a twig, the rustle of a leaf were no longer so significant to humans.

Industrialization was even more conducive to the cultivation of long-range views, first for the development of new products and then for the extension of markets. Although industrialization caused a great deal of environmental damage, the present awareness of this danger is leading to the formulation of long-range policies for the protection of the environment. Progressively also we have come to realize that many human activities affect larger and larger areas of the planet, to such an extent that we have now reached a point where we consider many problems from a global perspective.

Thus while the biological aspects of the human species

have not changed significantly during the past fifty thousand years, human attitudes have been constantly modified by the evolution of our relationships with the planet. Stone Age people related almost exclusively to their immediate surroundings, whereas today we begin to have the whole planet in mind and to be concerned with its distant future even when we engage in local action. We are becoming planetized probably almost as fast as the planet is becoming humanized, both processes being greatly accelerated by the increase in world population and by technological development.

No animal species is likely to have an awareness of the planet as a whole. Human awareness may have emerged gradually because, long before historical times, *Homo sapiens* had spread over most of the globe and established extensive trade routes that linked different population groups living in very different kinds of places. Memories of faraway places could be transmitted from one generation to another, and from one group to another, first through spoken, then through written language. It is certain that concern for the welfare of the planet and the human species as a whole already existed when the principles of the great axial religions were formulated around the fifth century B.C.

THE role of Judeo-Christian teachings with regard to environmental problems came to be seen in a new light following a statement originally published in 1953 by the Japanese Zen Buddhist D. T. Suzuki and later reformulated by Lynn White in an article entitled "The Historical Roots of our Ecological Crisis." The general theme of this article, which has become fashionable even and perhaps especially among theologians, is that the damage done to the planet by human intervention has its origin in Genesis 1:28, where man is given dominion over Nature. This hypothesis seems to me so completely at odds with historical facts that I have read more than 100 articles and books on the subject by lay scholars and theologians to better understand its meaning and implications. The range of this survey is illustrated in the Bibliography. I shall not review or discuss these documents but only present some of the conclusions that they seem to warrant.

1. As mentioned earlier, extensive and lasting environmental degradation occurred long ago in many places where people had not had any contact with biblical teachings—in many cases indeed long before the biblical writings.

2. Ruination of the land was usually the consequence of deforestation, exploitative agriculture, and ignorance of the long-range consequences of farming practices.

3. The early teachers of the Judeo-Christian-Moslem religions expressed great concern for the quality of Nature and advocated practices for its maintenance. In ancient Judaism, for example, it was a religious commandment to take fields out of cultivation every seventh year, a practice of great ecological value since it helped to maintain the fertility of the soil.

Similar recommendations can be found in the writings of many Christian and Moslem teachers. Although I have not consulted Oriental religious writings on this subject, the references to them that I have read reveal an attitude toward Nature very similar to that of Judeo-Christian-Moslem writers.

I hasten to acknowledge that the professed ideals of a culture, like those of its politicians, are rarely translated into actual practice, but this is at least as true of Oriental as of

Western peoples. China was one of the countries that experienced the worst deforestation and erosion, despite the professed reverence of Chinese scholars, artists, and poets for wilderness sceneries and for nature in general. Similarly, the teachings of St. Francis of Assisi probably had little if any influence on the destruction of wildlife by Italians and other Europeans.

4. People destroyed more of Nature as the centuries went by, not because they had lost respect for it, but because the world population increased and also because technological means of intervention became more and more powerful. The plains Indians lived in "harmony with Nature" as long as their impact in the hunt was limited to what they could do with bows and arrows, but they decimated the herds of bison once they could hunt them from horseback with firearms. As long as the Caucasians had only axes of stone and of metal, it took them centuries and even millennia to destroy a large percentage of their forests, but power equipment now makes it possible to clearcut immense areas in a very few years.

The historical origin of the present ecological crisis is therefore not in Genesis 1:28 but in the failure of people to anticipate the long-range consequences of their activities—consequences that have recently been aggravated by the power and misuse of modern technology.

Wise people have long realized that Nature must be treated gently and indeed with love. This awareness was translated into many traditional constraints on social conduct and into religious precepts that derived from a concern for conservation.

Ancient documents reveal that several countries of prescientific Europe had formulated strict regulations for the management of bodies of water, agricultural lands, and especially woodlands. A large percentage of the European forest has thus been managed according to precise official directives for many centuries. The religious organizations of medieval Christianity also did much to convert the inhospitable, swampy, tangled woodlands of northern Europe into agricultural lands that are still highly fertile today. Thus good methods of husbandry had been developed for sylviculture and agriculture long before the word *ecology* was introduced into scientific terminology.

In the United States, broad concepts of conservation reached the general public through the writings of George Perkins Marsh (1801–1882), also before the word *ecology* became fashionable. This American philologist, diplomat, and politician had no professional training in the natural sciences, but while serving as ambassador to Italy he realized that much farmland in Mediterranean countries had been damaged by poor agricultural practices and especially by deforestation. He presented his thesis in *Man and Nature, or Physical Geography as Modified by Human Action,* published in 1864, which he revised in 1874 under the title *The Earth as Modified by Human Action.* In it he attributed the decline of Mediterranean countries to the erosion, floods, and climatic changes caused by the cutting down of trees on the watersheds of rivers.

Marsh believed that the spongelike qualities of land under a primeval forest regulated stream flow, and for this reason he advocated wilderness preservation. Although he is justifiably regarded as a patron saint of the environmental movement, his initial motivation was not concern for the wilderness or a desire to identify humankind with nature, but the very anthropocentric thought that forested areas and wetlands play an important role in agriculture and are indeed essential to the economic health of agricultural nations. He was not opposed to modifications of nature; for example, he favored the planting of trees, where none existed naturally, and even such stupendous technological undertakings as the draining of the Zuider Zee.

The nineteenth-century conservation movement had its origin not in preservation of the wilderness for its own sake, but in purely human values. For aesthetic reasons, French painters of the Barbizon school convinced the authorities of the forest of Fontainebleau in 1837 not to cut down trees in some of the oldest and most picturesque parts of the forest. In 1853 negotiations between foresters and painters resulted in the establishment of 624 hectares *réserve forestière.* Continued negotiations between the two groups led to a master plan for the whole forest of Fontainebleau in 1864. The visual arts contributed to nature preservation also in the United States. In the late 1830s the painter of Indian life George Catlin (1796–1872) suggested that a great strip of the plains country be placed under government protection. This suggestion was not followed, but the dramatic landscape paintings of the Far West by Albert B. Bierstadt and Mathew Brady's photographs certainly contributed to the formation of the national parks. In 1863, Bierstadt visited Yosemite Valley, where he made studies for one of his most famous paintings. Yosemite became a state park in 1864 and Yellowstone the first national park in 1872. Writers, too, made their contribution to preservation. In the 1842 edition of his *Guide to the Lakes,* Wordsworth discussed landscape preservation in England and suggested that the Lake District be made national property. In 1844 another poet, William Cullen Bryant, called for an urban park in New York, a proposal that eventually led to the creation of Central Park in the late 1850s.

In the United States, national parks such as Yellowstone and Yosemite were placed under government protection not to save Nature in the wild but because their scenic and recreational values made them ideal for the entertainment of the public. When Gifford Pinchot created the National Forest system, which now manages 92 million acres of commercial forests, he wanted to create "forest *reserves*" (the original name of the system) to be used against the likelihood of a future "timber famine." Pinchot saw these forests not as a museum but as a place where the best forest-management practices would prevail. John Muir was at that time the only prominent advocate of wilderness preservation for its own sake, an attitude that did not begin to gain momentum until the middle of the twentieth century.

It would be out of place to describe here the history of the environmental movement, but a few facts illustrate the change in public attitude during the past century and the extent to which concern for the protection of nature is linked to concern for the quality of human life.

While a member of the Senate of the United States, Daniel Webster spoke to oppose the use of public funds for the construction of a railroad to the Pacific coast.

> Mr. President, what do we want with this vast worthless area? . . . this region of savages and wild beasts, of deserts, whirlwinds . . . of dust and prairie dogs? To what use could we ever hope to put these great deserts or those endless mountain ranges, impenetrable and covered to their base with eternal snow? What can we ever hope to do with a western coast of 3000 miles, rock bound, cheerless, uninvit-

> ing and not a harbor on it? Mr. President, I will never vote one cent from the public treasury to place the Pacific coast one inch nearer Boston than it now is.

Daniel Webster was wrong about the future of California, but he was expressing an attitude that was almost universal in his time—namely, that the wilderness is of no interest if it cannot be used to some practical end.

During the twentieth century, the ever-increasing membership of the Sierra Club, the Audubon Society, Friends of the Earth, and numerous other national and international organizations for the protection of the environment attests to the widespread public interest in the various aspects of Nature. This interest furthermore has many different facets. For some people it represents genuine concern for Nature itself in its various manifestations. Other people in contrast want to protect Nature because they know that its quality affects their own life.

During the 1960s public concern for environmental quality reached such a high level of intensity that it led to the organization of an international conference under the auspices of the United Nations. The focus of this conference, which was held in Stockholm in 1972, was not Nature but the dangers posed to human life by environmental pollution and by the depletion of natural resources. The official name of the 1972 Stockholm meeting, "International Conference on the *Human* Environment" [italics mine], clearly reflected the anthropocentric orientation of its organizers and participants. But there is now evidence for a broader approach to the protection of nature.

In 1973, the U.S. Congress enacted a law for the protection of endangered species. This topic came back for discussion in the Senate during the summer of 1978, because a dam planned by the Tennessee Valley Authority threatened to bring about the extinction of the snail darter, a rare species of a small fish having no known utility to humans. Authorization for the dam was voted down in the Senate by a surprisingly large majority of 94 to 3. Even more surprising were the sophisticated ecological statements made by one of the senators, Patrick Leahy of Vermont, in defense of the fish: "Ultimately, we are the endangered species. *Homo sapiens* is perceived to stand at the top of the pyramid of life but the pinnacle is a precarious station. . . . We exist on this orbiting globe locked and joined with the environment. We share the planetary gene pool with that snail darter in the Little Tennessee River."*

These words, and the vote in the Senate for the protection of the snail darter, reflect an attitude toward Nature that would have been inconceivable a few decades ago when permission was given to dam the Hetch Hetchy River in Yosemite. They also stand in sharp contrast, indeed comical contrast, to the speech in which Daniel Webster expressed, also in the Senate, his contempt for California as a worthless piece of wilderness.

After completing this chapter, I discovered by accident that when President Carter referred to "the moral equivalent of war" in his attempt to mobilize the nation against the squandering of energy, he had borrowed the phrase from the American philosopher William James, who had first used it in 1910, but in a very different context. For James, the equivalent of war was the struggle against Nature. He called for,

*Since this paragraph was written, the Senate has reversed its vote and approved (by the small majority of 48-44) the building of the dam.

instead of military service, a "conscription of the whole youthful population to form a certain number of years a part of the army enlisted against *Nature*" [italics his]. Thus James perceived Nature as a villain, armed with droughts, floods, hurricanes, and plagues. He would have found it difficult to believe that a time would come when the Senate would devote a session to the fate of a little obscure fish, and when the equivalent of war would be to protect Nature instead of protecting humankind from its dangers.

CHAPTER SIX

The Management of Earth

Improving on Nature

THE belief that we can manage the Earth and improve on Nature is probably the ultimate expression of human conceit, but it has deep roots in the past and is almost universal. The manifestations of this conceit can be recognized in the Stone Age people who domesticated animals and plants some ten thousand years ago; in the farmers of all ages who created agricultural land by cutting down the primeval forest, draining the marshes, or irrigating the deserts; in the planners of all historical periods who have converted natural landscapes and waterscapes into artificial parks and gardens; in today's homeowners who maintain lawns where brush and trees would naturally grow.*

*Since writing this chapter, I have discovered that belief in the possibility of improving on Nature was widespread among conservationists at the turn of the century. This belief is expressed repeatedly in the fifteen volumes of John Burroughs's writings on nature. In his Ph.D. thesis (George Washington University, 1979), Michael James Lacey demonstrates, furthermore, that improving on nature was considered an essential activity by the professional environmentalists of the

Human interventions into Nature have often been destructive. Many of them, however, have revealed potentialities of the Earth that would have remained unexpressed in the state of wilderness. We can improve on nature to the extent that we can identify these unexpressed potentialities and can make them come to life by modifying environments, thus increasing the diversity of the Earth and making it a more desirable place for human life. Before developing this theme, I must acknowledge that the term "improving on nature" would be nonsensical if I believed as most people do, that "Nature knows best." I shall therefore state my reasons for questioning the validity of this so-called ecological law.

Does Nature Really Know Best?

WHEN left undisturbed, all environments tend toward a state of equilibrium generally called the climax or mature state. Under natural conditions, most (but not all) wastes from living things are constantly recycled in the ecosystem, which becomes thereby more or less self-perpetuating. In a forest, for example, acorns are eaten by squirrels, which in turn may be eaten by predators. The products and dead bodies of plants and animals are decomposed by microbes that return the elements to the soil, enriching it thereby in humus and

late nineteenth century. In May 1908, President Theodore Roosevelt called a conference of governors at the White House to formulate rules for the management of forests and waterways. In his opening speech, he discussed renewable resources and stated that "man can improve on nature by compelling the resources to renew and even reconstruct themselves in such manner as to serve increasingly beneficial uses."

mineral nutrients. More vegetation grows out of these recycled materials, assuring the maintenance of the ecosystem.

The expression "Nature knows best" is justified when applied to such natural ecosystems, but is almost meaningless. There are no problems in Nature, only solutions, because the natural state is an adaptive state that generates a coherent system. This does not prove, however, that the existing adaptive state is the only possible one, or the best one. "Nature knows best" is the twentieth-century equivalent of Pangloss's affirmation in the eighteenth century that everything is for the best in the best of all possible worlds. The interplay between humankind and the Earth has often generated ecosystems that, from many points of view, are more interesting and more creative than those occurring in the state of wilderness.

Many natural systems, in fact, represent clumsy solutions to ecological problems, even when Nature has been left undisturbed. That the wisdom of Nature is often shortsighted is illustrated by the massive number of deaths—the so-called population crashes—that occur repeatedly in certain animal species, notably among lemmings and other rodents. The crashes are at best ineffective ways of reestablishing equilibrium between population size and local resources.

Nature also fails in many cases to complete the recycling processes considered distinguishing characteristics of ecological equilibrium. Examples of such failures are the accumulation of coal, oil, shale, peat, gas, and other fossil fuels of organic origin. These fuels are largely derived from the bodies of plants and other living things that have become chemically stabilized after undergoing partial decomposition. The fact that they have accumulated in fantastic amounts obviously means that they have not been recycled.

Paradoxically, people help in the completion of the cycle when they burn the fuels and thus make the carbon and other constituents once more available for plant growth.

The accumulation of guano provides another example of recycling failure on the part of Nature. This material, now used as a fertilizer, consists of the excrement deposited by birds for centuries or even millennia on certain islands and cliffs. For example, millions of seabirds use the Chincha Islands off the coast of Peru as a resting place and breeding ground; their droppings have formed layers of guano from 60 to 100 feet in thickness. Guano is rich in compounds of nitrogen, phosphorous, and potassium, and its accumulation therefore means recycling failure. Here again, people complete the recycling process by collecting guano and spreading it as agricultural fertilizer on fields, where it reenters the biological cycle in the form of plant nutrient.

I have mentioned population crashes and recycling failures because these examples have been singled out for criticism by Professor Don Goldman in the thoughtful review he wrote of my article "Humanizing the Earth." Professor Goldman considers that the population crashes among lemmings and other animals are good techniques of population control since "the lemming survives, which means that the species (with its peculiar population oscillations) has survival value and occupies a niche." I can only answer that, having watched animals during periods of crash, I am convinced that they suffer before dying. I cannot agree with Professor Goldman when he writes that this is a matter of indifference to them. Population crashes are a clumsy method of population control.

Professor Goldman also feels that "it is perhaps too hasty to say that petroleum under the ground and guano on an island are failures on nature's part. Everything *does* go somewhere but maybe not fast enough to suit man." Admittedly, the accumulation of guano may play some long-range role in the order of things, but since no one has suggested what it could possibly be, I still consider that using guano as fertilizer and thus returning its components to the natural cycles is a better ecological solution than letting it accumulate on rocks.

Finally, Professor Goldman states that, in considering the human creations of landscapes, "we must distinguish between *mere* furniture moving and true creativity. In gardens and parks, man has *simply* taken those elements that nature has provided and rearranged them to his taste" [italics mine]. In my opinion, these words do not do justice to the thought and genius that designers and planners have displayed in creating styles of landscape architecture. I can only restate my view that the diversity of the natural world has often been enriched by human management. What Professor Goldman calls "furniture moving" is what humankind adds to natural processes and therefore constitutes an essential part of civilization.

Professor Goldman does not disagree with my statement that "the human use of natural resources and of technology is compatible with ecological health, and can indeed bring out potentialities of the earth that remain unexpressed in the state of wilderness," but he feels that I should have added a number of qualifying phrases, such as ". . . *if* the number of humans are kept within the carrying capacity of the space they occupy or . . . *if* the technological demands the humans impose are within the environment's capacity to absorb and heal or . . . *if* humans purposely [*sic*] adopt a standard of living based on sustained yield." I did not feel it necessary to restate once more these obvious conditions. People who read my writings have long been aware of the finitude of the Earth and do not need to be told that there are inescapable limits to quantitative growth.

In part to reassure Professor Goldman, I have nevertheless stated in several chapters of this book and in two appendices some of the dangers that I consider most threatening to the welfare of humankind and our planet. However, although the avoidance of dangers may help in survival, it is not sufficient for creative evolution.

The Human Origin of Many "Natural" Environments

In the last chapter of *A Sand County Almanac,* Aldo Leopold made a statement that provides the philosophical basis for the "land ethic" now identified with his name. "We must quit thinking about decent land-use as solely an economic problem. Examine each question in terms of what is ethically and esthetically right, as well as what is economically expedient. *A thing is right when it tends to preserve the integrity, stability, and beauty of the biotic community. It is wrong when it tends otherwise*" [italics mine]. This statement is often interpreted to mean that Leopold was opposed to the transformation of Nature by human activities, but this is not the case.

Leopold naturally pleaded for concern with ecological health but he did not state anywhere in his book that the natural biotic communities of the wilderness are the most desirable and that it is wrong in principle to alter them. In fact, he described several ecosystems of which he obviously approved, even though they were the products of accidental or intentional human interventions. For example: "In the case of Kentucky, we do not even know where the bluegrass came from—whether it is a native species or a stowaway from Europe." We know only that "the cane lands when subjected to the particular mixture of forces represented by the cow, plow, fire and axe of the pioneer, became bluegrass."

In northeastern Europe, the

> land physiology remains largely normal despite centuries of human occupation. As to Western Europe, its flora and fauna are now far different from what they were two thousand years ago. . . . Yet the soil is still there and, with the help of imported nutrients, still fertile; the waters flow normally; the new structure seems to function and to persist. There is no visible stoppage or derangement of the circuit.
>
> Western Europe, then, has a resistant biota. Its inner processes are tough, elastic, resistant to strain. No matter how violent the alterations, the pyramid, so far, has developed some new *modus vivendi* which preserves its habitability for man, and for most of the other natives.
>
> Japan seems to present another instance of radical conversion without disorganization. . . .
>
> Health is the capacity of the land for self-renewal. Conservation is our effort to understand and preserve this capacity.

Although Leopold deplored the fact that many land formations have been damaged by mismanagement, he did not claim that this is an inevitable consequence of human intervention into Nature.

In many parts of the world that have been occupied for long periods of time, human beings have deliberately created artificial ecosystems that can be regarded as the analogues of natural communities. In Asia and Europe, particularly, large areas have been patiently shaped through a trial-and-error process similar to organic evolution. Early settlers have thus created artificial ecosystems possessing high levels of ecological diversity and stability in areas where the original ecosystem of the wilderness has been essentially destroyed.

Ecological diversity has been increased, of course, by the deliberate or accidental importation of animal and plant species. Even the honeybee, which is native to Europe, Asia, and Africa, was introduced into America around the year 1600 and more recently into Australia and New Zealand. Today it prospers in all parts of the temperate zone.

Hawaii provides a striking example of increase in ecological diversity through the introduction of foreign species. Before the white settlements, the Hawaiian biota was relatively simple and disharmonic. It was deficient in terrestrial vertebrates; it had no pine trees, no oaks, no maples, no willows, no fig trees, no mangroves, one single species of palm, and only a few insignificant orchids. Needless to say, countless plant species have now been introduced into Hawaii and other Pacific islands, where they have prospered even better than in their native habitats.

Even more interesting than the introduction of plants and animals from one part of the world to another is the fact that ecological diversity can be modified and indeed can be increased by deliberate changes in the environment. Human analogues of natural communities have thus emerged in the replacement of the wilderness by a multiplicity of new microhabitats. There is such a wealth of successful artificial ecosystems throughout the world that it suffices to mention only a few types, selected because they correspond to sharply different climatic regions.

Most of the hedgerows that line the country roads in England and continental Europe and also separate the fields are completely artificial ecosystems dating from the very early Middle Ages or even before. The patchwork of small fields and hedges that was in the past the typical landscape of East Anglia was laid down much later, having been established as a result of the Enclosure Acts. Whether of ancient origin or dating from Enclosure time, European hedgerows now constitute an ecosystem of great diversity and charm. Unlike many American hedges, they are not rows of trimmed shrubs of a single species, but complex populations of trees, shrubs, flowering plants, grasses, small mammals, songbirds, and a host of invertebrates. They serve as reservoirs for animal and plant species that would not prosper in the primeval forest or in a completely cleared landscape. In other words, they have an ecological diversity of their own.

Hedges also contribute in many other ways to the quality of the landscape, for example, by providing habitats for spiders and other natural enemies of insects that prey on crops; by acting as windbreaks and thus protecting the land; and by providing shade for domestic animals and hikers.

In Japan, substantial hedges commonly surrounding farmhouses serve as windbreaks and provide a decorative element. Their Japanese name, *ikegaki,* means "living curtain," a reflection of the fact that they harbor a great diversity of living forms.

In the lowlands of the south Asian tropics, where population densities have long been extremely high, successful artificial ecosystems have been developed that closely inte-

artificial ecosystems have been developed that closely integrate numerous types of animal and plant life in such a manner as to achieve an extremely high efficiency in the use of space and nutrients. The wet-rice ecosystem of southern China is as remarkable for its complexity and stability as for its productivity. The rice paddies are usually located in valleys that can be subjected to controlled flooding. The water may be first used to irrigate vegetable gardens that it fertilizes with its organic matter; it produces a fine bloom of algae that in turn supports various forms of aquatic animal life. Oysters and other shellfish are raised in it and it sustains shoals of fin fish. The ecosystem is kept in smooth operation by natural processes of fertilization and pest control. Chemical pesticides are not used or needed because pigs, chickens, ducks, and frogs feed on the refuse; these animals become themselves succulent parts of the human diet. Weeds that are missed in the course of this cleaning process are pulled up and used as mulch or as feed for pigs and fish. All nutrients including animal and human dung are recycled into the system.

Each animal species of the wet-rice ecosystem has its own niche. Some species of carp are surface eaters, others grass eaters, and still others are plankton feeders for different water levels; the mullet acts as a bottom detritus feeder. This completely artificial ecosystem is probably as stable as any in the world and more productive than most with regard to both quantity and diversity of food.

Some extremely "unnatural" ecosystems were created in the arid regions of the Middle East in the very distant past. Three thousand years ago, the Persians brought water from the mountains to the arid plains on a scale rivaling that of the great Roman aqueducts. In parts of the Negev where only desert scrub grows naturally, one finds remnants of long-abandoned farmsteads and other evidence of prolonged and intense human occupation.

Flint implements cover huge areas in the central Negev plateau, from the coarse tools of primitive peoples to delicately shaped arrowheads and boring tools of the end of the Stone Age. Copper and early Bronze Age sites are found along the course of the Nahal Habesor. Abraham and his successors operated a regular donkey caravan service between Mesopotamia and Egypt and relied upon supplies of fodder and facilities along the way, part of which crossed the Negev. The most intensive settlement of the Negev took place under the Nabataeans, Romans, and Byzantines, who secured the ancient trade routes with fortresses that eventually grew into full-fledged cities. They built ingenious water conservation systems to utilize every drop of scanty rainwater for the irrigation of vegetable plots, orchards, and vineyards. This period of activity and civilization, lasting nearly a thousand years, came to an end with the Arab conquests of the seventh century, and until recent decades only roaming Bedouin tribes derived a precarious livelihood from the barren landscape.

During the past few decades the Israelis have once more made the Negev a prosperous agricultural region. By methods very similar to those used two thousand years ago, the few millimeters of rain that constitute the total annual precipitation are channeled into microcatchments for fruit trees, food crops, and medicinal plants. Large yields could thus be obtained even under the extremely dry conditions that prevailed during the 1962–63 season.

In the Arab world, certain natural ecosystems have been replaced by almost completely artificial ones. Although many of them have undergone degradation for lack of care, others have been more successful. It is largely through human intervention that the Nile valley has long been profoundly different from the neighboring desert. The Ghuta orchards of Damascus, the palm groves of Matmata in Tunis, the artificial oases of the Maghreb are examples of productive and pleasant ecosystems created and maintained by human effort under arid conditions.

In many places where abundant water is available, immense areas of desert have been recently converted into rich agricultural lands. This process, which has been termed *dedesertification,* seems to have been particularly successful in the Imperial and Coachella valleys of the United States, in the "Hungry Desert" of Uzbekistan, and in the "90-Mile Desert" of Australia. There are many places, however, in which more than water is needed to convert the desert into agricultural land.

In Australia the soil of the 90-Mile Desert east of Adelaide contained virtually no phosphorus, copper, zinc, or nitrogen, and supported only a miserable vegetation. During the 1940s these missing elements were added to the soil, which was then seeded with subterranean clover *(Trifolium subterraneum)* and grasses. In twenty years, the low-nutrient desert had become a rich pastureland. Although the 90-Mile Desert is perhaps the most outstanding example of how nutrient-poor soils can be improved, there are other cases in Australia and elsewhere. In northern Germany and the northern Netherlands, for example, the almost unproductive heaths are now thriving farmlands, thanks to the scientifically controlled addition of proper nutrients. Both in the Australian deserts and in the heaths of northern Europe, introduction of the missing mineral nutrients resulted not only in the lush development of economically important plants, especially grasses, but also in the virtual disappearance of the plants of the original ecosystem.

The Arab countries of the Middle East are now making plans for the greening of the desert. Although these countries are wallowing in petrodollars, their leaders realize that their underground wealth will virtually disappear within a few decades if exploited at the present rate. To prepare for the future, they are planning to invest income presently derived from oil in the development of human and material resources that will remain productive after the oil wells have dried up. Some Arab countries have initiated programs aimed at achieving technological self-reliance through the creation of industries and scientific institutes. Studies are also under way to determine what kinds of crops and livestock suitable to semidesertic conditions can be improved by selection; giant irrigation resources (in part from desalinated water) are being considered in the hope of converting some 10 million acres of sand into agricultural land.

On the other hand, it has been suggested that a sounder policy might be to create self-supporting cities in the desert designed to grow their own food, for example, on the roofs of buildings where greenhouses could be built. The economy of these hypothetical desert cities would be based not on scarce and unreliable water supplies but on the abundant sunlight that can be used to produce solar energy for the development of intensive agriculture and certain specialized industries. By concentrating agriculture and industry in limited areas with fairly high population density, most of the desert could be kept as wilderness—a natural resource that will be increasingly valuable on the overpopulated Earth.

In all the examples mentioned, the creation of agricultural

lands has or will have been achieved through modifications of existing ecosystems, but there are situations in which it can almost be said that the soil itself is of human origin. There are approximately 8 billion acres of potentially arable land on Earth, but the soil of half this area is too poor to be of use in agriculture. As we have seen, however, the addition of some mineral elements may be sufficient to render the soil fertile. From observations that he made in western Europe at the end of the last century, Russian sociologist Peter Kropotkin came to the conclusion that *any* kind of soil could be made to yield large crops if it were properly treated. As an extreme example, he mentioned the fantastically high fertility of the Paris truck farms, in which "the soil is always made, whatever it originally may have been. Consequently it is now a usual stipulation of the renting contracts of the Paris *maraîchers* that the gardener may carry away his soil down to a certain depth, when he quits his tenancy." Another example of successful farming under unfavorable conditions is the use as soil of the hardened mica stored in the lagoons created by the exploitation of China clay in Cornwall; with proper fertilization the mica proves to be excellent for the production of potatoes, other vegetables, and fruits. Needless to say, large investments of time and capital are needed for converting such unpromising materials into fertile agricultural soil, but this may become increasingly feasible as more is learned of plant physiology.

The case of the Netherlands will be considered at some length for two reasons. On the one hand, this country is the most spectacular and seemingly the most successful example of land creation in the world. On the other hand, it provides dramatic illustrations of the dangers inherent in even the most sophisticated management of the Earth.

Six out of ten people in the Netherlands today live on land below sea level. In the western and northern parts of the country, much of the land has been reclaimed from submerged areas and must constantly be protected against water from rivers as well as from the North Sea. For the past fifty years, nearly 2 percent of the national income has been spent on dredging, draining, and reclaiming. Yet despite this huge environmental expense, the Netherlands is one of the healthiest and most prosperous of countries, with the highest population density in the world. The saying "God created the world but the Dutch made Holland" seems to justify the human conceit that we can indeed improve on Nature.

The modifications of Nature that have shaped the Netherlands date from the very beginning of the Christian era when settlers built artificial earth mounds to protect themselves against floods and storm tides. As early as the eighth or ninth century, Leyden, Middleburg, and similar towns on the west coast had been established on mounds measuring several hundred feet in diameter and fifty feet in height—veritable hills in this flat landscape. In our own time, Amsterdam's western dormitory towns have also been built on ground artificially raised, in this case by using subsoil dug from the Storeplas polder; the excavation site has become a lake and the rich polder topsoil has been used to create artificial lawns, parks, and recreation grounds.

The building of dikes began in the eighth century. By the end of the thirteenth century, many parts of the coastal belt were protected against the sea water, but the protection was never completely effective. Time and time again, villages and vast areas of farmland were devastated by river floods and sea water that had breached the dikes. The repair and replacement of the dikes has long been one of the dominant preoccupations of Dutch life.

Another constant preoccupation has been the reclaiming of farmland from submerged areas, especially in the polders. The general procedure is to dig a wide canal around a body of fresh or salt water and to use the soil removed from it for the construction of a dike surrounding the future polder. Water is then pumped from the polder area and discharged into the canal, which carries it to a river or the sea. Meadows and crops are established as soon as the polder soil is sufficiently dry.

Pumping out the water is naturally a crucial part of all the processes involved in land protection and reclamation. The pumps used to be powered by windmills, the thatched roofs and turning sails of which became a characteristic endearing feature of Netherlands scenery. Windmills, however, have now been replaced by electric pumps. This change made possible one of the greatest technological achievements of the twentieth century—the draining of the Zuider Zee.

The Zuider Zee used to be a very shallow body of sea water silted with rich alluvions from the several rivers that discharged themselves into that part of the North Sea. Early in the twentieth century, a dam was built across the mouth of the Zuider Zee to seal it off from the sea. The dam, completed in the early 1930s, is twenty miles long; it carries a wide motorway and has locks for the passage of ships. Sluices and a large pumping station are used to regulate the water level inside the dam. The drainage operations have now converted the area into four dry polders and a freshwater lake. Some 850 square miles of land have thus been reclaimed from the sea and are used for agriculture or housing. The fresh water from the inner lake serves as drinking water for human settlements and for irrigation in periods of drought.

The catastrophic storm tides that occurred in February 1953 led to an extension of the dike program to seal off the Scheldt and Rhine estuaries and shorten the southern shoreline. The salt coastline of the Netherlands, which was 1,230 miles in 1840, will be reduced to 420 miles after completion of the system of dams, sluices, storm barrages, and powerful dikes along the vulnerable coastline of the Rhine-Maas-Scheldt delta.

Rotterdam and its harbor is another spectacular example of the transformation of the Netherlands by technology. Although the city was almost completely destroyed during the German air raid of 1940, its harbor is now said to be the largest, busiest, and most modern in the world. The new section, suitably called Europort because it serves practically all Europe, was created by gigantic dredging operations that deepened to seventy-five feet over a length of seven miles a shallow natural water link to the sea, now called the New Waterway.

Paradoxically, the greatest dangers facing the Netherlands today come from the very success of the management of nature. Despite constant improvements in design and operation, the dikes cannot be made completely safe against the ferocious storms of the North Sea, let alone against human malice. The northwest gales of February 1, 1953, pushed the waves over the dikes and smashed through them in many places. Close to 800 miles of dikes were damaged and the sea water covered 400,000 acres of villages and farmland. Some 2,000 persons died; the repair of the dikes cost 6 percent of the national budget, and it took seven years before agricultural production returned to normal. As population, urbanization, and industrialization continue to increase in the Netherlands, natural disasters are likely to become even more destructive of human lives and economic resources. And many other difficulties can be anticipated even without catastrophic storms. In Rotterdam, the air and water are increasingly polluted by oil refineries, chemical factories, and spillage from oil tankers. The deeper the New Waterway is dredged, the more salt coming from the sea contaminates the drinking water and the agricultural Westland. More important perhaps in the long run is that the ever-increasing technological management of Nature inevitably decreases the intimacy between people and things that used to be such an appealing aspect of life in the Netherlands, as expressed by Rembrandt, Vermeer, and other Dutch masters.

The Management of Woodlands

MANY forests of western Europe have been carefully managed for several centuries and look very different from what they were in the state of wilderness. In Germany, for example, the most typical areas of the famous Black Forest consist chiefly of Norway spruce, but only because German foresters prevent beech trees from becoming established and covering the land as they would if nature were allowed to follow its course. In vast areas of the American Pacific northwest, similarly, the forest is managed to keep the Douglas fir dominant. In England, the so-called New Forest has been under constant management since 1079—900 years ago! The trees are maintained under several different systems of sylviculture, creating a great diversity of biological habitats and scenery. Among the most famous areas are the unenclosed woodlands known as "the Ancient and Ornamental," which consist of very old beech and oak trees, with "an understorey of holly, surrounded by open parkland, grassed lawns, open heath, bogs and thickets of birch containing holly, oak, occasional ash." Parts of the New Forest consist of woodlands and broad heaths grazed by the ponies and cattle of farmers holding ancient grazing rights. Other parts are used as recreational grounds.

Deforestation is still one of the major threats to the global ecosystem, but several reforestation programs have been initiated, not only on the American and European continents but also in the People's Republic of China, South Korea, the state of Gujarat in India, and several parts of Africa. I shall mention here only a few cases of reforestation, selected because of their interest or picturesque value.

The forestry departments of African governments have been producing millions of tree seedlings for distribution to villages in the hope of encouraging communities to reforest, but with little success. An outstanding exception is Ethiopia, where some 200,000 acres of eucalyptus have been planted. Ironically, most of this vast rural afforestation was accomplished by illiterate peasants using nonscientific methods, even before a professional forestry service had been established. The program seems to have begun around 1905, and its success in the Addis Ababa region seems to have played a role in making this city the permanent capital of Ethiopia.

Scattered specimens of different species of large trees indicate that some trial-and-error experimentation was carried out before the decision was made as to the species best adapted to local conditions and agricultural practices. At present Ethiopian farmers know and plant only two species, *Eucalyptus globalus* at altitudes above 2,000 meters, and *Eucalyptus calmaldulensis* at lower altitudes. Furthermore, *E. calmaldulensis* is grown not only under arid conditions to which it is best adapted but also in areas of high annual rainfall. Although the two species chosen are not the most productive, they are superior to others with regard to adaptability and in particular they possess a toughness that enables them to survive the casual planting techniques used by Ethiopian farmers.

Within the past twenty years considerable areas of level Ethiopian farmland have been reforested with eucalyptus because farmers have found that they can earn more from the trees than from wheat and barley crops. This area was once covered with great forests but had been progressively denuded and eroded by crop farming and cattle grazing; thus a process of land degradation three thousand years old is now being reversed.

SUCCESSFUL examples of afforestation by more orthodox methods can be found in many different parts of the world. Consider a number of examples.

Monterey pines were planted more than 150 years ago as a windbreak around the Scilly Isles off the west coast of Cornwall. They grew so well that they now permit the cultivation of early vegetables and daffodils for the London market.

Oaks and alders grow vigorously in Amsterdam's Bos Park on land reclaimed from the sea.

Trees are taking hold in the Palestine desert, especially in the Israeli kibbutzim. South of Esdraelon Valley, the hills were planted during the 1930s with Aleppo pines not native to the region that are now more than fifty feet high and provide shelter and food for a diverse fauna and flora. Wildflowers, birds, deer, and foxes prosper in association with the Aleppo pine. The exclusive use of these pines, however, is objectionable for both ecological and aesthetic reasons. The new plantings in other parts of Israel therefore include a more diversified selection of trees chosen not only on the basis of botanic and economic criteria but also because their shapes are better suited to the local landscape. Moreover, certain areas have been left bare of trees so as not to mask the architectural beauty of the rock formations.

In the southwest of France, the region known as Les Landes, covered with pine trees until the beginning of the Christian era, provided large amounts of resin to Rome. In 407 A.D., vandal hordes swept through the region, razed the villages, dispersed the population, and set fire to the pine forest, thus completely destroying the vegetation cover of a vast sandy area. Prevailing winds from the west began to move the sand from the ocean dunes. In its movement eastward this sand covered farms and villages, dammed streams, and formed marshes. Two centuries ago the region was an immense marshland of some 2.25 million acres, with scant vegetation but rich in mosquitoes. Malaria was prevalent and the sickly population was the poorest in France.

In 1798, Napoleon appointed one of his most famous engineers, Brémontier, to control the movement of the sand from the dunes. This was achieved by establishing a protective littoral dune. Later some 1,600 miles of ditches were dug to carry off the surface water to streams or lakes, and some 2 million acres were planted chiefly in pines, with some oaks. This artificial forest has been managed ever since its creation, more than a century ago, and has proved constantly productive of a great variety of valuable products derived from the trees, in particular resin from the pines, cork from the oaks, and of course a variety of timber. The region that used to be desolate and extremely poor is now one of the most prosperous in France and in addition can boast of many health resorts.

The Landes region is now generally assumed to be natural and its forest is taken for granted by the French public. François Mauriac, one of France's most famous contemporary novelists, born and raised in this region, once wrote, "The Landes have not changed. They will never change. . . . The innumerable pines deprive the eye of any horizon, compelling it to search for a narrow opening of sky between their soaring tops."* Mauriac did not mention that these pines are not native to the Landes and that they could be cultivated there only because the swamp had been drained.

San Francisco's Presidio Park, one of America's most beloved sites, also illustrates that the creation of artificial woodland can result in successful landscapes. The extensive wooded areas of the Presidio contrast sharply with the bareness of the surrounding northern tip of the San Francisco peninsula, but the woodland is not a gift of nature. The Presidio was treeless until the implementation of a detailed "Plan for the Cultivation of Trees upon the Presidio Reservation," presented in 1883. Its author, Major W. A. Jones, recommended that "Young trees should be fenced off to keep out grazing cattle," and that "growth [be] pretty much restricted to eucalyptus and evergreens because deciduous wouldn't grow—too windy."

*Les Landes n'ont pas changé. Elles ne changeront jamais. . . . Les pins innombrables frustrent l'œil de tout horizon, l'obligent à chercher un ciel étroit entre leurs cimes vertigineuses." *La Province.*

Some 55,000 acacia pine, cypress, and eucalyptus trees and 5,000 native redwood, spruce, and madrone trees were planted. For some period of time, at the invitation of the army, native San Franciscans joined in Arbor Day celebrations and helped to plant trees in the Presidio. The actual number of trees planted is not known, but the planting continues. In addition, there is a large amount of natural seeding, especially pine and cypress. Of the Presidio's 1,400 acres, approximately 280 are today defined as woodland. But as in the case of Les Landes, few people know that the area was once barren and that it became one of the most attractive parts of San Francisco only through human effort and ingenuity.

Artificial Environments from the Industrial Wilderness

In the western hills some six miles from Peking are still to be found the waterways, island groves, and hills of the famed Yuan Ming Yuan, or Garden of Perfect Brightness. This enormous complex of landscapes, waterscapes, palaces, and smaller residences was created by the Manchu emperors in the first half of the seventeenth century. These were two principal units, the Old and the New Summer Palaces, the so-called palaces consisting in reality of numerous self-contained idealized living units so distributed among the valleys, hills, and lakes that the many branches of the emperor's family could each have peace and privacy. Both palaces were destroyed by the British in 1860. The New Summer Palace has been restored and now extends over 823 acres of artificial scenery, four-fifths of which are water.

The islands, mainland forms, and even the hills of the

Peking Summer Palace are entirely built from excavation in a marshy floodplain. They had been originally planted with species introduced from the farthest reaches of the then known world, and these contributed still more to the artificiality of the landscape.

Eighteenth-century European travelers were enormously impressed by the "natural" charm and beauty of the complexes and by "all the Bridges and all the Groves . . . planted to separate and screen the different Palaces and to prevent the inhabitants of them from being overlooked by one another." They did not seem to have realized that the Chinese landscape architects had achieved this "natural and wild view of the country" by creating a completely artificial environment out of the natural marshy floodplain.

The park of Versailles was established shortly after the Peking Summer Palaces, also on marshy land, but it was designed to fit the concept of nature and the social patterns that prevailed in France during the reign of Louis XIV. Whereas the Chinese landscape architects had tried to symbolize the mysterious complexities of nature and had emphasized privacy, Le Nôtre composed in Versailles a linear design that could be apprehended immediately and totally and that was in tune with the public life of the French court. In Versailles as well as near Peking, the result was achieved by a complete transformation of the natural environment by armies of workers and mechanical equipment. Some twenty thousand workers were simultaneously employed on the Versailles site during critical periods of the park's development.

It is ironic that the completely artificial garden paradise of the Manchu emperors was one of the main inspirations for the "return to nature" movement in European landscape design. Many artificial landscapes and waterscapes have since been manufactured to convey a natural atmosphere. In our own time, the Peking Summer Palace has provided the prototype for the new Chicago Horticultural Society Botanic Garden now being developed north of the Skokie marshes on an abandoned farm and on other adjacent degraded areas. Some three hundred acres are being used to create a complex of artificial streams, lakes, islands, and hills from exhausted agricultural land and from grossly polluted drainage channels.

Construction of the earthwork and garden grounds of the Chicago Botanic Garden was started in 1966. Three years later, the land sculpturing had been completed, the lake basins and streams filled, and the new landscapes seeded. Horticultural collections from many parts of the world are being assembled to create a garden second to none, as had been the ambition of the Manchu emperors in the Peking Summer Palaces.

If the landscapes and waterscapes of the Chicago Botanic Garden continue to develop successfully, this will justify the hope that wastelands unfit for human use and providing only marginal habitats for animal and plant life can be converted into settings of ecological interest and visual beauty. Simple observation reveals that environments seemingly as unpromising as railway sidings actually permit the spontaneous development of many types of wildflowers, including those that contributed so much to the spectacular aspect of the American tall-grass and short-grass prairies. Moreover, the seclusion provided by railroad embankments permits large colonies of flowers to build up and thus to produce large masses of colors. Rubbish heaps, worked-out quarries, and the sides of abandoned canals are among the many other types of degraded environments that are rapidly occupied by a diverse flora and fauna. During World War II, unexpected types of vegetation commonly appeared on bombed urban sites. The number of species of plants and animals that appeared without any human help was surprising. Within a few years even trees grew up among the rubble and provided cover for insects and birds, including rarities like the black redstart, several pairs of which nested and reared their young within the boundaries of the City of London.

In addition to these cases of the spontaneous return of various living forms on degraded areas, there are many examples of deliberate and successful attempts to use degraded areas for the creation of artificial ecosystems. New York City's Central Park, for example, was developed on what was then a desolate area at the northern edge of the city. Much of the Brooklyn Botanic Garden was established on a site once used as an ash dump by New York City; the area was rehabilitated by a very few workers and a team of horses at the turn of the century.

In Paris, the Parc des Buttes Chaumond (a residential district) was created about 1863 from abandoned limestone quarries. According to a description by an Englishman in 1877, enormous old quarries surrounded by acres of rubbish once occupied this spot. By cutting away the ground around three sides of these quarries and leaving the highest and most picturesque side intact, stalactite caves sixty feet high have been constructed. Enormous curtains of ivy drape the great rock walls with the most refreshing verdure at all seasons.

In London, a three-acre site that used to be a desolate parking lot just across the Thames from the Tower of London has now become the William Curtis Ecological Park. The rubble has been removed and soil has been added and carefully landscaped with undulations and valleys. A good-sized pool has been constructed in marshy ground. Although the ecological park has some areas left for spontaneous regeneration, most of it is being carefully managed. Many native trees and shrubs have been planted and wildflower seeds have been scattered. "The pond is already colonised by insects and invertebrates one might not expect to find in such a location. A kestrel nests in the one remaining large tree, all that is left from the rural countryside built over in the last century."

In Poland, a park of 600 hectares has been built by the population of the Silesian coal fields on an area that used to be covered with heaps of cinders that harbored only a sparse ashy vegetation. As judged from photographs, this polluted and badly degraded land has now been converted into beautiful woodlands, with streams and lakes and a great diversity of recreational facilities.

The need to restore areas damaged by mining and industry is increasingly leading these industries to create essentially contrived environments. In Australia, for example, it has recently become the practice to remove the most vulnerable plant species from the land over titanium mines and to place them in nurseries. The topsoil is saved. As soon as the excavation is completed, the topsoil is put back, fertilizer is added, and the plant species that had been maintained in nurseries are replanted. Near Pittsburgh, a five-acre strip-mined coal pit has been transformed into a stock game preserve. Layers of processed municipal wastes were sandwiched between layers of soil and compacted to prevent spontaneous combustion before being covered with fertile soil. In Indiana, recreational facilities have been created in a state park on land profoundly disturbed by strip mining. Compacted trash has been used to create ski slopes in flat country; a sanitary hill known as Mount Trashmore has been created at Virginia Beach. Several similar programs for the reclaiming of mined areas and the use of wastes have been conducted in Great Britain, the United States, and Germany, and appear to have been successful. A longer time will be needed of course to evaluate the final outcome of the reclaiming process.

The most spectacular example of reclaiming of open mines is probably that of the "lignite" mine near Cologne, West Germany, known under the name Fortuna. The mine is a thousand feet deep, occupies a surface of approximately fifteen square kilometers, and has been operated for thirty-five years. As the mine advances, agricultural fields, villages, and roads are completely destroyed but rebuilt almost immediately. To this end, the topsoil of loess is set aside before the mining operations are begun. For each ton of lignite mined after the removal of the topsoil, there are two tons of useless soil, which is also set aside and then returned immediately to the hole. The topsoil is replaced on the surface and enriched with fertilizers, then planted with trees and crops. The fields are said to achieve a normal state within approximately five years. Since 1964, 3,000 hectares have thus been reconstructed with lakes, beaches, forests, amusement grounds, roads, villages, and even a canal. Agricultural fields have been created, and more than 60 million trees have been planted. According to the overall plan, 12,000 hectares will eventually be divided into 4,000 hectares for agriculture, 700 hectares for forty-five lakes (of which seven will be suitable for sailing), 5,000 hectares for trees, and the balance of 2,300 hectares for villages and roads.

An even more artificial landscape is being created by the Badische Anilin und Soda Fabrik (BASF) at Ludwigschafen, also in West Germany. Every year BASF produces 300,000 tons of wastes that are disposed of as artificial hills created on an eighty-hectare area located on Flotzgrün Island, forty kilometers upstream in the Rhine. The area is first covered with a thick layer of lime, then the wastes are compacted and piled upon the lime. When the hill of wastes has reached the proper height, it is covered with a thick film of plastic in order to prevent the rain from dissolving its toxic components and carrying them into the water table. Then the plastic film is covered with a thick layer of topsoil that is fertilized and planted with a varied vegetation. The process began in 1967 and one can already see a green hill with trees of fair size dominating the Rhine. Four million tons of wastes have thus already been used; from two to four thousand tons are brought in every day and there is room on the island for 21 million tons.

THE two most famous artificial environments of the eighteenth century, the summer palaces of the Manchu emperors and the park of Versailles, were located a very few miles from the then largest cities in the world, Peking and Paris. Yet land was so plentiful at that time that immense and essentially nonproductive landscapes and waterscapes could be created out of these marshes. All the other artificial environments just mentioned were built from degraded sites and from wastes—the new kind of wilderness created by the Industrial Revolution. Just as shortages of raw materials now make it necessary to recycle resources, so will shortages of natural environments make it necessary to reclaim the industrial wilderness—the millions of acres of stripped, eroded, and otherwise degraded lands; the thousands of miles of rivers and other bodies of water contaminated with acids, metals, and organic poisons; the countless extraction pits and sanitary fills. Reclamation will not be cheap. According to a recent estimate, more than $250 billion would be needed to repair the damage done by mining to the land in Appalachia alone. But since land in decent condition and pure water are now in short supply, reclamation is inevitable. Fortunately, it is possible in most cases. The recycling of degraded environments is one of the most urgent tasks of our age.

Even the most successful programs of reclamation and the best artificial environments cannot of course duplicate the subtleties and complexities of natural environments; but most of them will improve with time. In the course of their long existence, many artificial environments such as the olive groves of the Mediterranean region, the wet-rice ecosystems of southern Asia, and the hedgerow country of East Anglia have spontaneously acquired an astonishing ecological complexity and human quality. It is encouraging to remember also that, as revealed by pictures of the time, the trees of the great eighteenth-century English parks were puny when first planted, and that the grounds developed their present subtlety and richness only after many decades. The new Chicago Botanic Garden looks immature today and will need much time to reach the splendor imagined by its architects. There is every reason to believe, however, that the area will progressively mellow visually and acquire a great diversity of living forms. In most parts of the Earth, time and life modify artificial environments, conferring on them the poetic nobility of natural creations.

CHAPTER SEVEN

Of Places, Parks, and Human Nature

Place versus Environment

THE word *environment* does not convey the quality of the relationships that humankind can ideally establish with the Earth. Its widespread use points in fact to the present poverty of these relationships. In common parlance, as well as etymologically, the environment consists of things around us, out there, that act on us and on which we act. Whether good or bad, the physical components of our surroundings are foreign to us and we are foreign to them.

We expect more of the environment in which we live, however, than conditions suitable for our health, resources to run the economic machine, and whatever is meant by good ecological conditions. We want to experience the sensory, emotional, and spiritual satisfactions that can be obtained only from an intimate interplay, indeed from an identification with the places in which we live. This interplay and identification generate the spirit of place. The environment acquires the attributes of a *place* through the fusion of the natural and the human order. All human beings have approximately the same fundamental needs for biologic and economic welfare, but the many different expressions of humanness can be satisfied only in particular places.

Places providing satisfactions that transcended biologic and economic requirements already existed in the Stone Age. Cave paintings in France and Spain, the circles of Stonehenge and Avebury in England, the alignments of Carnac in France, the statues of Easter Island and other megalithic structures required a large share of the resources of Stone Age communities, yet they did not serve any obvious biological or practical purpose. These stupendous creations gave a special human meaning to the places where they were established. Subsequent societies have also devoted much imagination and labor to the creation of places that had no biological or economic value—whether Tibetan monasteries or Roman arches of triumph, Greek temples or Gothic cathedrals. In many cases, admittedly, the great creations of the past were erected by slaves or other forced labor, but not always—witness the building of Chartres cathedral, which involved the entire local population, including the rich and powerful.

The importance of place, above and beyond the quality of the environment, explains why a sense of nostalgia suffuses attitudes toward the land. Europeans commonly yearn for the type of places they associate with the great cultural periods of their respective countries, as evidenced by the design of their parks and gardens and by the stories they tell of the pleasures of life in historic landscapes. Americans have of course also been exposed to cultural conditioning, but their nostalgia for the land generally has a more recent origin. Although there is a long history of human occupation in North America, the Indians left few permanent structures on the land and had only little influence on the social patterns created by European settlers. American nostalgia toward the land is largely based on views that the settlers derived on the one hand from fanciful descriptions by Europeans who imagined the New World as a pastoral Arcadia, and on the other hand from the actual experiences of the immigrants who had to struggle with the wilderness.

The first mental images of America among Europeans came from descriptions popularized in Europe during the seventeenth and eighteenth centuries. After the discovery of the New World, European writers imagined that the whole continent was a pastoral paradise, gentle in climate and in sceneries. This image was not entirely false but was derived chiefly from romanticized accounts written by explorers who had spent short periods of time in a few semitropical regions of North, Central, and South America. Some of the early settlers in the Southern colonies did manage to create cultural environments that fit this utopian picture. The wilderness had little appeal for Southern patricians and their families, who preferred instead polished surroundings in which to cultivate gracious life-styles. William Byrd, owner of Westover plantation in Virginia, wrote for example that "a Library, a Garden, a Grove, a Purling Stream" provided the most suitable conditions for a happy life. Jefferson's Monticello was the embodiment of this eighteenth-century ideal, a far cry from the wilderness. Today's landscapes, especially the grounds around homes, still reflect a romantic longing for the pastoral life-style of the pre–Civil War period, even though this style did not last long and was always limited to a few regions and people.

For most early settlers, as well as for the people who later moved from the Atlantic to the Pacific coast, the perception of America was very different. For them, life commonly implied struggle with the wilderness, an experience dreaded by most immigrants, who were overwhelmed by the sheer size and wildness of environments that were unlike anything they had known in Europe. Awed and frightened in particular by the primeval forest, they immediately cut down trees wherever they settled, not only for the sake of the timber or to create farmlands but also to establish their homesteads in open areas that gave them a greater feeling of security. The settlers' struggle with the trees is the heroic phase of American history and probably contributes the deepest element to national nostalgia. The ordeals it implied account for the American tendency to identify the word *nature* with a forested wilderness and to associate moral and manly values with life in the wild.

Nostalgia for the wilderness still expresses itself in the way Americans manage the land. French and Italian landscape architects generally emphasize formal designs; the English aim at a more natural but nevertheless well-controlled appearance in their parks and gardens. North Americans, in contrast, commonly favor a more casual treatment of public grounds and try to maintain in them a somewhat rough and unpruned atmosphere, as if to pretend that the places in which they live are still close to the wilderness.

The Emergence of Places

THE English hedgerow or moor countryside, the European bocage, the Mediterranean hill towns, the Pennsylvania Dutch country, the Chinese mountain and water landscapes call to mind ecosystems intimately associated with certain ways of life. The fitness they exhibit between local people and nature entitles them to be called places, in the sense I have given to this word. The catalyst that converts an environment into a place is the process of experiencing it deeply —not as a thing but as a living organism. Fitness is achieved only after slow progressive reciprocal adaptations and therefore requires a certain stability of relationships between persons, societies, and places.

Most places have emerged spontaneously, or at least without conscious design, to meet some human need. Before modern times, for example, the various landscapes of the Earth represented patterns of organization that had evolved empirically from local agricultural practices and ways of life. In most parts of Europe, farmers live even now in compact villages, the reason being often the difficulty of obtaining adequate water supplies; they travel back and forth from the village to their fields and pastures. Farms tend to be isolated only where water is readily obtained from shallow wells, the farm buildings and the farmer's home being then located among his fields and pastures. In such areas, villages serve chiefly as commercial and social centers.

Wherever farm buildings are assembled in compact villages, at least in continental Europe, roads tend to be focused on the village center and also to go from village center to village center. Where farms are isolated, in contrast, the roads that link them are supplemented by others that constitute a regional network. Before the seventeenth century in England, most agricultural communities were self-sufficient; a circular road system around the village facilitated the movements appropriate to a fairly standardized rotation of crops and cattle. When land patterns changed after the Enclosure Acts, it became more practical to use roads leading from the human settlements into the fields. Thus different kinds of agricultural settlements generate different road patterns that contribute to the spirit of place.

Each urban agglomeration also soon acquired a spirit of its own. It used to be thought that cities had developed from earlier agricultural villages that had simply grown larger with time. This view has been made untenable, however, by the discovery that a number of human settlements reached fairly large size, perhaps close to 10,000 in population, before the development of agriculture. Such was the case for Jericho in the Jordan Valley, Tell Mureybit in Syria, Shanidar in Iraq, and Catal Hüyûk in central Turkey.

The preagricultural urban settlements probably emerged as centers of exchange for special kinds of goods, many of which were traded over very large distances. Among these goods were copper, lapis lazuli, and especially obsidian, a volcanic glass much prized in the ancient world for the manufacture of mirrors and other household objects. Ancient cities may also have been the first sites of pottery making and of metallurgic industries. Being primarily centers of exchange, they were located on trade routes, commonly on harbors or rivers. Even today, practically all important cities are located on or near large bodies of water because they initially achieved their greatness through water links to distant parts of the world.

From its very beginning, each ancient city acquired an individual pattern of organization that persisted, at least in its center, throughout its subsequent growth. An obvious influence was physical location, since buildings had to be fitted to the topographical features of the site—around a harbor, along the shores of a river, on the slopes of a hill. Practical considerations may have led to the gridiron system of streets that became prevalent very early. It was used as far back as five thousand years ago in Egypt and in the Indian city of Mohenjo-Daro. It persisted in the Roman castrum, the medieval eastern German towns, and the bastides of southern France, in certain European cities of the seventeenth and eighteenth centuries, and of course in the modern cities of North America.

Even more important than these practical criteria for the design of cities were the habits that governed the lives of their inhabitants. Cities very soon became centers of religious worship; indeed, many of them may have started as such. Their pattern of physical and social organization usually expressed symbolically the attitudes of the people toward their gods. Thus a concentric design emerged early from the psychological desire to place the monuments dedicated to the local deity on an elevated point and to arrange human activities around and below this holy site, in a well-defined hierarchical order. Places of worship such as the temple or the cathedral became the *axis mundi.*

Concentric or stellar plans naturally evolved in the course of history to fit the spirit of the time. In Versailles the monarch of divine right replaced God as the focal point of planning. The industrial-commercial revolution of the nineteenth century created new practical needs that did not fit the concentric plan, but the age-old feeling that human environments should be centered continued to influence urban design long after religious faiths had lost their original power. Some recent garden cities, as for example those envisioned by Ebenezer Howard for England, and the Israeli kibbutzim, are seemingly based on the assumption that concentric living promotes happiness or at least social cohesion. Whereas the cathedral and the town hall used to be at the urban center, modern settlements such as the Swedish satellite cities or Sun City in Arizona tend to be organized in relation to shopping malls.

Social criteria inevitably had to be considered in all early planning, even the most unconscious, especially since cities were centers for the exchange of goods, information, and ideas, and soon became sites of wealth and power. Fear influenced the choice of topographical location and the creation of structures for defense against potential enemies. Social prestige required that prominent locations be reserved for the buildings and grounds of the ruling social classes. Markets and the convenience of merchants made it necessary to open wide streets and other communication systems that overshadowed the original geomorphic and concentric town plans. The need for public areas where the community could assemble for discussions or celebrations probably gave rise to open spaces such as the Greek agora, the Italian piazza, the French mall. Aztec towns, which seem to have been roughly rectangular, also had a central plaza.

All these religious and social criteria constrained further development within the city itself and compelled many people to move to the outskirts. In certain countries, furthermore, regulations forced traveling foreign merchants to stay outside the city walls and therefore to develop for themselves a new kind of cluster settlement. In the Near East, only the local retail trade was conducted in the bazaar streets within the walls, whereas huge market centers became established outside. This centrifugal movement has now been intensified by the creation of satellite cities and by the urban sprawl into suburbia and exurbia.

THE techniques and materials used for building differed greatly in the past according to local resources and climatic conditions. Early peoples learned to build from branches, lumber, stone, adobe, brick, or whatever other material was locally available and was suitable for the climate—including snow in the polar regions and gypsum in the Sahara, where this substance is readily available. Even under the most trying conditions, people managed to build adequate shelters not only by using local techniques and materials, but also by inventing an immense variety of designs. As one travels from the cold and bleak countries to the warm and sunny lands, the shape, slope, and dimensions of the roofs in the old settlements change according to the amount of snow, rain, and insolation to be expected in the locality. Porches, patios, architectural and landscaping styles, even the orientation of the streets and the design of public spaces all reflect the climatic and topographic peculiarities of the different regions. People who built their own shelters were alert to the shortcomings of design and progressively introduced suitable corrections. The continuous feedback between building techniques and fitness for living led everywhere to an architecture without architects, which gives to each of the old human settlements a distinctive local character and contributes to the genius of the place.

Economic imperatives, however, are now taking precedence over the religious, social, and climatic criteria that generated many types of successful human settlements under a great diversity of environmental conditions. Houses of worship and princely mansions are dwarfed today by gigantic business establishments. The same building materials, techniques, and styles are used everywhere, irrespective of climate and topography. Instead of being organized according to a hierarchical pattern, churches, houses, schools, offices, shops, and factories are now strung almost indifferently along seemingly endless roadways, thus generating linear cities. It has been further predicted that these linear cities will in turn be united in a continuous, anonymous ecumenopolis.

The longing for variety, however, may act as a force countering the trend toward uniformity. Constantinos Doxiadis, the Greek planner who coined the word *ecumenopolis* and thus publicized the concept of the continuous world city, later expressed his concern for the loss in the quality of life caused by the standardization of human settlements. To convey his ideal of a more intimate relationship between humankind and the local environment, he introduced the word *entopia* (from the Greek, meaning "in place"), denoting a type of planning and building that takes advantage of local conditions and thus is in harmony with nature.

The skeptics of course have good reason to believe that *entopia* is an abstraction that cannot come to reality in the technological and standardized atmosphere of present-day civilization. But the optimist can answer that diversity continues to prevail over much of the world and keeps alive the spirit of individual place.

Even in areas that have been plowed, bulldozed, built upon, there persist some stubborn peculiarities of the place that affirm its identity: a spatial and textural relationship between rocks, water, soil, and slopes, despite rough handling by real estate developers; a characteristic luminosity or poetic melancholy of the air despite the smog; an overpowering mood of the seasons that air conditioning cannot eliminate; the continuing evidence of historic landmarks even where monuments have been destroyed. It is the persistence of these characteristics that intrigues people who return to their village, city, region, and that gives them the thrill of instant recognition. Whether the thrill be one of pleasure, bittersweet nostalgia, or overwhelming sadness, the visitor senses that the physical location is not only an environment but a human place as well.

Landscaping for Human Nature

EACH region and each community thus has its own spirit of place resulting from the prolonged interplay between people and their surroundings. This process of reciprocal adaptation occurs in ordinary life through continuous minor changes in the people and their environment, but a more conscious process of design must have presided very early over the development of parks and gardens. Whether these artificial grounds were created chiefly for private or public enjoyment, they could be successful only if they were ecologically viable and also satisfied instinctive needs that human nature has derived from its evolutionary past.

In *Modern Painters,* John Ruskin pointed out that "every Homeric scenery intended to be beautiful includes a grove, a meadow and a fountain." This applies to the composition of successful landscape paintings. Claude Lorrain, for example, commonly balances long, open vistas with deep shadows under massive canopies of foliage, a composition that endows his paintings with a feeling of security and tranquility. In fact, most painters who have tried to depict scenes of happiness out of doors have placed them in a clearing but near a wooded area, usually with a stream, fountain, or other body of water nearby.

The various styles of landscape architecture also provide a sense of both open space and refuge with many elements of visual diversity, but they express in addition the characteristic attitudes of the society from which they emerge. People living in flat areas, for example, commonly build artificial mounds, either as burial places or simply to make their mark on the landscape and to dominate their surroundings. In Sumer, the largest surviving monument of this type is a ziggurat known as "Hill of Heaven," which dates from about 2250 B.C. and is approximately eighty feet high. Other ziggurats of the same period probably reached more than 100 feet in height and must have looked like natural rocky eminences, because some of their terraces were planted with trees. This practice was perhaps the origin of Semiramis's Hanging Gardens of Babylon of the fifth century B.C., one of the seven wonders of the world.

The creation of artificial hills has remained ever since a feature of grand-scale landscape architecture, as seen in the Peking Summer Palace built in the seventeenth century, in the English parks of the eighteenth century, and in the new Chicago Botanic Garden of our own time. The desire to occupy an elevated position dominating the landscape probably had a biological origin and eventually resulted in an aesthetic experience satisfied by artificial hills with no practical utility. As is frequently the case, a desire of biological origin eventually evolves into a sociocultural attitude. Persepolis, perhaps the most splendid human settlement of antiquity, was located on a huge podium thrusting majestically outward from the mountains and dominating the plains. Later in the Western world, noblemen's castles and robber barons' mansions were more often than not located on hills because their physical elevation came to be identified with social dominance and distinction.

Geometrical arrangements are extremely rare in nature and indeed practically nonexistent either in natural topography or living forms. Yet rectangular or square gardens and parks occur in the very first civilizations, perhaps because human eyes and minds were early conditioned by the regular patterns of irrigated fields between the Tigris and Euphrates.

Many tales of antiquity refer to special localized places where humans enjoyed fruit trees, shaded groves, soft grass, sweet-smelling flowers, song birds, friendly animals, and other riches of nature. Such an idyllic place called paradise in Persian was geometrical and was commonly separated by walls from the desert or other nearby wilderness. In the Bible, Adam and Eve lived in a garden similar to the Persian paradise before being driven out of it into the wilderness.

Most ancient Eastern and Mediterranean countries seem to have shared the same ideal of an orderly humanized landscape. The earliest records concern Egyptian gardens from about 2000 B.C. and Babylon's Hanging Gardens. Although these early achievements of landscaping have disappeared, the drawings of them that have survived give an idea of their geometrical sophistication. The best surviving gardens of the Middle Eastern type are found in Marrakesh, Morocco, and in El Alhambra, the Granada's Generalife, and the cathedral garden of Seville in southern Spain, where they were created during the Moslem occupation.

The first hunting parks, which came soon after the domestication of the horse by the Assyrians, were also laid out geometrically with trees often imported from afar, as was much of their population of wild animals. For some five thousand years, at least since Sumerian times, gardens and parks have thus been designed to express an idealized view of natural and agricultural scenes. This trend reached its most extreme form with Le Nôtre's park at Versailles, where nature was completely managed.

The classical park tradition of Europe started in imperial Rome, but a simpler version has its origin in the Christian Church. Monks, especially of the Benedictine order, created gardens in their monasteries and gradually extended cultivated lands outside the monastery walls. A complex of orchards, vegetable gardens, herb gardens, flowering meadows, springs and streams with grassy banks, and well-tended groves eventually became the northern counterpart of the Persian paradise. For medieval Christians, gardens and parks created out of the wilderness were places in which to experience on Earth some of the joys associated in the public mind with a heavenly paradise. The Roman villa garden as described by Pliny was a complex pattern of terraces, slopes, flights of steps, and flower beds landscaped to form a *natura architecturalis.* With time, the landscape element became more and more dominant in the villa gardens and finally produced the gardens of the Italian Renaissance.

French landscape architecture began in the sixteenth century with extensive sections for vegetables, fruit trees, and flowers outside the castles. It developed its national style when architects learned to achieve a formal unity between buildings and grounds. Flower beds were extended beneath windows like ornate carpets and long avenues of trees led into open nature. Groves, flower beds, and shrubberies became places for social intercourse. Management of nature and formality of design probably reached their highest levels in the Park of Versailles. This corresponded well to the ritualization of French court life under Louis XIV but was not incompatible with more informal attitudes. Everyone who was "reasonably" dressed was admitted to the park, where countrymen's carts jostled those of noblemen.

Although the French baroque style of landscaping was rapidly imitated in all the European courts, including that of England, it was already being criticized at the end of the seventeenth century. The longing for nature expressed by Jean Jacques Rousseau and by numerous English writers and artists led to a style of landscaping that aimed at being true to nature but was in reality an attempt to make artificial landscapes of human design look as if they were natural. It is ironic that the English landscape architects of the back-to-nature movement in the eighteenth century derived much of their inspiration from two very artificial sources—on the one hand, the paintings in which Claude Lorrain and Nicholas Poussin had idealized Italian sceneries; on the other, as mentioned in Chapter Four, the Chinese parks where large landscapes and waterscapes had been literally built and dug according to a certain Oriental philosophy of nature.*

For more than three thousand years, we have been elabo-

*The longest continued tradition of sophisticated parks originated in China. It began before the Christian era and reached its peak with the construction of the very artificial Old and Summer Palaces near Peking. From China, this tradition passed in the sixth century into Japan, where it has continued to be practiced and to evolve.

rating on Homer's description in the Odyssey of Alcinous's garden, which was close to woodland but open on one side, with two fountains amid plantations of fig trees, olive trees, and vineyards. When considered from the point of view of their use in a park, however, the various components of the vegetation have a meaning different from what they would have in agricultural land. This is apparent in the use of the words *grove* and *lawn,* which do not refer to trees or grasses as botanical structures or as plants of practical utility, but rather to settings where human sentiments about nature are most appropriately experienced or expressed.

All types of landscape styles—Italian, baroque, romantic, "natural," Chinese, or combinations thereof—came to be practiced in the United States, but the chief American contribution to landscape architecture was derived from the western wilderness. Yosemite, Yellowstone, the Garden of the Gods, and many other parts of the Rockies and Sierras caught the attention of early explorers not as areas of wilderness per se but because they provided on a grand scale and *au naturel* the kind of physical atmosphere that the various styles of landscape architecture had tried to create artificially. Yosemite and Yellowstone were officially called parks, not wilderness areas, precisely because they appeared to have been designed by Nature for the admiration and the entertainment of the people.

It seems odd at first sight to use the same word *park* for the artificial creations of landscape architects and for certain regions that are naturally picturesque. Yet this usage is justified because *park* refers above all to a kind of landscape that satisfies fundamental visual needs acquired in the savanna-type country of the Stone Age. The parks of the wilderness, however, added something new to human life by providing an opportunity for perceptions and values not readily experienced in the parks of civilization. They help humankind to rediscover and appreciate some aspects of physical and human nature that civilization has blurred or rejected altogether.

Gardens and parks differ profoundly from the wilderness in the way they are developed and maintained, even when attempts are made to make them look natural. Good landscape architects have always known that they must work with Nature, not because their creations are similar to Nature's but because of the necessity to respect ecological constraints. This is just as true of parks in the baroque style as of those in the so-called natural style. The grounds were swampy at Versailles where Le Nôtre used the water to create his famous *nappes d'eau;* he designed the majestic straight *allées* from existing woodlands by taking advantage of the fact that the area was essentially flat. The grounds and bodies of water of New York City's Central Park now look natural but are in fact almost completely artificial. Frederick Law Olmsted, who drew up the plans for the park in 1858, loved the English pastoral scene and admired in particular the Birkenhead estate, where large amounts of earth had been moved in 1847 to create artificial hills, valleys, and serpentine paths with a belt of detached terraces and villas around the park proper. Memories of Birkenhead and the English countryside were in Olmsted's mind when he designed Central Park with his associate Calvert Vaux, and created it by using the most advanced techniques of civil engineering, with some four thousand men at work on the grounds at one time. The final outcome was that Olmsted and Vaux imposed on the Central Park area a pattern that they, and their contemporaries, regarded as "natural" but that was in reality contrived. Landscape architects always impose a human pattern *on* nature, even when they make it a point to work *with* nature.

Parks have become more and more different from the wilderness with regard to the use that people make of them. To see but not to be seen was the formula for survival in the wilderness; elevated areas helped people to keep watch over their surroundings. For this reason, irrespective of styles, practically all parks possess physical structures, artificial if not natural, that provide opportunities for both refuge and prospect.

But early in their history, parks served as places where the chief goal was not to observe nature but rather to be seen by other human beings. People of low rank in the royal parks of the eighteenth century wanted to be seen by people of higher rank and if possible by the king. Marcel Proust's duchesses, demimondaines, and *jeunes filles en fleurs* took an afternoon drive in the Avenue des Acacias to be seen rather than to enjoy the fresh air and the greenery. On Rome's Pincio people of both sexes take a *passagiata* chiefly to watch each other. As to the highest hills in almost any park anywhere in the world, for one person who climbs it to gain a better view of the scenery, a hundred drive to the cocktail lounges and restaurants where the "best view" is that of other persons who are drinking, dining, or dancing, usually unaware of Nature except in its human forms.

Working with Nature has always been one of the fundamental imperatives of landscape architecture. Regardless of styles in design, whether in imperial Rome or in eighteenth-century England, planners have had to contend with natural constraints and to work with certain types of vegetation and topographical situations from which to create artificial landscapes and waterscapes. The genius of place is different in Vermont from what it is in New Mexico, and this is inevitably reflected in landscape design, whether for a public park or a private country club. Other requirements deriving from unchangeable human needs are compatible with a wide range of designs selected to fit terrain, social custom, and individual tastes, as interpreted by the imagination of each particular landscape architect.

The two complementary aspects of landscape architecture, the ecological invariants of a given area and the artistic imagination of a particular planner, can be read in two phrases that Horace Walpole wrote on two different occasions. In a letter giving an account of a visit to France, he stated that "they [the French] can never have as beautiful a landscape as ours, till they have as rotten a climate"—a witticism that clearly emphasized the role of ecological imperatives in landscape design. (Half a century earlier, the French Prince de Ligne had expressed a comparable opinion about English landscaping. Writing of Capability Brown's lawn-based parks, he remarked that the English had made splendid use of their climate to grow grass; "their verdure they owe to their fogs.") But Walpole also wrote elsewhere that "an open country is but a canvas on which a landscape might be designed," thus affirming that everywhere on Earth the human mind can recognize unexpressed potentialities and create from them new patterns compatible with the laws of nature.

CHAPTER EIGHT

Humankind and the Earth

Environmental Ambivalence

MOST of us are ambivalent about defining environmental quality. Our attitudes are governed more by habit than by logic; they put us either in league with nature or in conflict with it, depending upon our past experiences and what we mean by the words *conservation* and *preservation.*

On the one hand, we extol the virtues of the wilderness and want to preserve it intact for its own sake. The condors, the redwoods, the Far West canyons, the tropical rainforests and frozen tundras, the marine estuaries and other types of wetlands, the wildlife of the African savanna and of the Asian jungle, the whales, the porpoises, and the multifarious fauna and flora of the oceans may or may not be of any use to us, but this is irrelevant. We fight on their behalf because we believe in their right to existence and in their importance as unique forms of creation.

On the other hand, we resent any change in our environmental heritage, even though it was created out of the wilderness by human activities. We want to protect the farmlands of New England, the hedgerows of East Anglia and the European bocage, the canal systems that artificially link natural bodies of water, the rows of trees planted in linear patterns, the villages and monuments that were erected at the cost of much environmental damage. We struggle to save these human creations, forgetting that all of them represent areas deforested, swamps drained, hillsides gouged of their stones and sand. In brief, we want to save both the wildernesa and the environments that have been created by destroying the wilderness.

Ambivalence toward nature is not peculiar to our civilization. Oriental peoples express mystical attitudes toward wild mountains and waterfalls; their religions endow nature with a divine spirit. But the ancient Chinese destroyed much of their forests, the Japanese distort the trees in their temples, and Oriental peoples in general have created some of the world's most humanized landscapes—whether for the production of food, as in the mountain-water agricultural economy, or for the glory of gods and rulers, as in their countless temples and palaces. In ancient Persia, nature was admired in areas called paradise, which were in reality enclosed luxurious parks where animals were kept for human pleasure. When the classical authors of ancient Greece and Rome wrote about nature, they did not have in mind the grandiose wilderness that still existed at that time in the Mediterranean world, but orchards, olive groves, and cultivated fields.

In our time most people who extol the wilderness have little if any contact with it. The usual practice is to spend a short period in a wild area for the sake either of excitement or physical and mental well-being, then to write about the experience in a comfortable office or home. The wilderness readily lends itself to declamation, but this is rarely converted into consistent behavior. In fact, some of its most famous devotees have expressed doubts about their own faith.

Viewing the Maine forest from the top of Mount Katahdin, Thoreau, for example, was shocked by the contrast between what he saw and the kind of Nature he knew around Concord. The Maine scenery appeared to him "even more grim and wild than you had anticipated . . . savage and dreary." In it he felt "more lone than you can imagine." His writings make it clear that he was not interested in the wilderness per se but was instead in search of an inner experience of wildness. He could create this experience out of solitude in the civilized quietness of Walden Pond or in the fields and woodlands around Concord, but not when engulfed in the real wilderness. Lewis Mumford expressed a similar attitude in a recent interview: "I haven't ever gone off into the wilderness for any length of time. I haven't felt the inner need for that. . . . I've only walked a few miles on the Appalachian Trail, though I was [in] on its very inception. . . . But I do need solitude every day and some contact with nature in a more or less primitive form, if only in the form of pasture."

The wilderness can be considered merely as an ecosystem. From the human point of view, however, it helps us to recognize that a great deal of fundamental wildness still persists in our own nature. Thoreau's phrase "in wildness is the preservation of the world" truly expressed his philosophical attitude as well as that of Emerson and other New England transcendentalists, but it referred to a human experience of wildness and did not imply that the Earth must be maintained in a state of wilderness.

John Burroughs is another American nature writer who had ambivalent views about the wilderness. He elected to spend his adult years in a rugged section of Dutchess County, New York, along the Hudson River, where he lovingly observed and described the forms of wildlife around him, but there were many kinds of Nature he did not like. Whereas he greatly admired the humanized English scenery, he thought that the Badlands of Utah were "as new and red as butcher's meat" and that much of the American West looked like "the dumping ground of creation." The very last essay in his last book deals with the struggles and sufferings of wild animals in Nature and ends with the phrase, "The wild life about us is full of tragedies," indicating that he had no illusions about the ethical values of natural systems.

John Burroughs experienced the tropical environment in only one place, Jamaica, and his feelings about the tropical wilderness had a character of hostility very similar to those expressed by Aldous Huxley in his essay "Wordsworth and the Tropics" (see page 15). American scholar and naturalist Joseph Wood Krutch also had reservations concerning his need for the wilderness. He had abandoned New York City and lived in Arizona because of his love for the desertic Southwest. Eventually, however, he experienced in the desert "something like terror. . . . We may look at it as we look at the moon but we feel rejected. It is neither for us or for our kind." Throughout the ages, human beings have responded to the various forms of wilderness with both fascination and fright.

As pointed out earlier, there are good biological reasons for an ambivalent attitude toward the wilderness. We admire deserts, high mountains, and tropical rainforests and recognize their essential ecological importance, but we know that these environments are fundamentally alien to our biological nature and that we can function and survive in them only with the accoutrements of civilization.

Certain sociocultural reasons, furthermore, are making it increasingly difficult for us to accept the idea that areas once humanized should be allowed to return to a state of wilderness. Throughout the temperate zone, especially in North America and western Europe, large areas of agricultural land are being abandoned because they are no longer economically profitable for farming. Most of this abandoned farmland was forested before being used for agriculture, and trees spontaneously grow back on it as soon as farming is discontinued. In view of the widespread concern about the disastrous ecological consequences of deforestation, one might think that the return of trees on abandoned farmland would cause universal satisfaction, but this is not the case. Although we want to preserve the forest where it still exists, we tend to resent its return in areas where it has been cleared by human effort.

One obvious reason for objecting to the return of trees is that their presence would make it more difficult and costly to reopen the land for agriculture when this becomes necessary. As the population of the great food-producing states increases, California and Texas may eventually consume most of what they produce, and will therefore have little food to export. Moreover, the fear of labor conflicts and fuel shortages makes it dangerous to depend exclusively on long-haul sources of food supply. For these reasons, the states of Connecticut, Massachusetts, New Jersey, and New York have begun to explore means of preventing further deterioration of their agricultural lands. New policies of land use and tax incentives for the preservation of farmland are among the measures being considered to prevent the forest from reestablishing itself in areas that had been cleared for agriculture and that might once more be needed for food production in the future.

Probably more important than fear of food shortages is the fact that most people enjoy the visual and other sensual qualities of open rural landscapes. Allowing trees to grow back where they once existed seems to the public as bad a form of environmental degradation as cutting them down where they now exist. In fact, a time may come when industrialized societies will create financial incentives to enable farmers to continue providing not only crops of economic value but also the spectacle of cattle pasturing in meadows, hedges fluttering with song birds, and other bucolic scenes that we mistakenly regard as manifestations of undisturbed nature.

Preservation versus Management of the Wilderness

THE most paradoxical manifestation of environmental ambivalence is that we now consider it necessary to manage not only humanized environments but also the wilderness itself. Many forests of Europe and North America are under strict human management. We make ourselve judges of how many deer or elephants should be allowed to live in a wilderness park for their own good and for the good of their habitats. We increasingly interfere with natural forces that have created the wild flora and fauna.

The managers of parks and reserves in the United States used to extinguish any fires as fast as possible, whether caused by lightning or human carelessness. Smokey the Bear is still the symbol of this policy. Ecologists have pointed out, however, that fires play a useful role in certain forest ecosystems; they are essential not only to prevent the accumulation of brush and debris, but also to permit the germination of seeds of certain tree species and to facilitate the return of nutrients to the soil. For these reasons, it has now become official policy to let many natural fires follow their own course, at least under certain conditions, and even to start fires wherever they are thought to be useful.

Most paradoxical of all, there have been plans to protect Niagara Falls against the impact of natural forces. Niagara Falls has been receding for several thousand years. The water that insinuates itself into rock crevices freezes and expands in winter, thereby causing cracks and making debris fall and accumulate at the base of the falls. Erosion and rockfalls are part of the natural evolution of the system, but the accumulation of rocks and debris disturbs the ideal image of the falls held by the public. At the instigation of local citizens, both American and Canadian, a "Fallscape" committee was formed a few years ago to develop methods for controlling the falls in order to preserve their visual "purity." Although this project has been abandoned, it provides a telling example of the fact that the quality of wilderness is as much a state of mind as a physical reality.

The love of Nature thus frequently generates ambivalent attitudes regarding the comparative merits and rights of the wilderness and of humanized environments. One of the most celebrated confrontations to emerge from this ambivalence concerned the Hetch Hetchy Valley in Yosemite National Park.

Fewer than twenty-five years after the creation of the

park, the city of San Francisco in 1913 asked permission to build a dam across the Hetch Hetchy River to create a municipal water supply and a source of hydroelectric power. John Muir, who had been the prime mover in the establishment of Yosemite as a national park in 1890, was then president of the Sierra Club and led the opposition to the dam. "Our wild mountain parks," he stated, should be "saved from all sorts of commercialism and marks of man's work." Labeling his opponents "temple destroyers," he exclaimed, "Dam Hetch Hetchy! As well dam for water tanks the people's cathedrals and churches." However, many eminent conservationists were in favor of San Francisco's application. One of them was Theodore Roosevelt; another was Gifford Pinchot, who, as head of the United States Forest Service, had popularized the word *conservation* a few years earlier. Even some members of the San Francisco–based Sierra Club believed that damming the river for domestic use represented "its highest use"—surely as anthropocentric a statement as could ever be made. After a bitter conflict, Congress and the president approved the dam. The Hetch Hetchy Valley became a reservoir, even though many regarded it as scenically beautiful as the Yosemite Valley itself. John Muir died, heartbroken, a year after the loss of the Hetch Hetchy battle.

Since 1913 there have been other proposals for building dams and other utilitarian structures in the national parks, including the Grand Canyon, but all of them have so far been voted down. The increase in popularity of the wilderness, however, is creating a new kind of danger for the national parks and a new type of ambiguous attitude toward Nature. In Roderick Nash's words, "The gravest future threat to America's . . . wilderness will not come from its traditional enemies (the economic developers) but ironically indeed from its newly acquired friends." When the director of the Sierra Club selected as one of the club's mottoes, "Take nothing but pictures, leave nothing but footprints," he probably did not realize that the continued impact of footprints would eventually cause extensive damage to the trails of the national parks and other wilderness areas. The wilderness is being loved to death.

The conflict between preservation and recreation is becoming more intense as more people seek the wilderness experience. The number of visits to all the units of the National Park Service increased from 114 million in 1965 to 228 million in 1975, thus doubling in ten years. There were some 8,000 people per square mile in the Yosemite Valley on a recent Fourth of July weekend—all of them trying to escape from civilization. The popularity of the national parks is obviously making it difficult to preserve the wilderness itself and to maintain the quality of the experience for visitors. Trails and vegetation are damaged, the animal population is disturbed, and intimate contact with nature becomes increasingly rare. Crowds cannot enjoy solitude.

The situation is not peculiar to the United States and probably reaches its extreme in Japan. The Fujisan mountain in the Fuji-Hakone-Isu National Park has long been venerated as a religious cultural object and serves now in addition as an escape from urban life. Seventy million people visit the area each year and 1 million climb to the top of the 12,467-foot volcanic cone of Fuji. In the summer months, 25,000 people make the ascent *per day:* "The lines of climbers wind up the switchbacks like huge, multicolored snakes. At night, with flares, they resemble glowworms." Even the so-called roadless wilderness areas of Japan are thronged; there are long waits at most of the cliffs suitable for rock climbing. Helicopters regularly haul away trash left by tourists in the alpine region. The Japanese are so used to crowding in their daily life that perhaps they are able to experience the wilderness even in the sardinelike setting of Fuji. Perhaps in Japan people find it possible to achieve "solitude" even in multitudes.

The United States, however, still holds the record when it comes to creating a vulgar circus atmosphere in its national parks. Colored spotlights have been focused on the famous Yellowstone Park geyser, Old Faithful, and recorded music is played between its hourly eruptions to ease the impatience of tourists. The "fireball" at Yosemite, a huge bonfire created from interwoven inflammable materials, is pushed over a three-thousand-foot cliff every night for the entertainment of tourists assembled on the valley floor below. Luxury hotels, ski lifts, tennis courts, swimming pools, and golf courses are located on the park grounds; visitors are entertained with staged amusements such as the regular feeding of bears with hotel garbage. Universal Studios, a subsidiary of the Music Corporation of America (MCA), operates thirteen liquor stores within Yosemite. The firm was allowed to film a television series in the park and to paint some of its cliffs. Conventions held in Yosemite hotels run by MCA at times prevent real tourists from finding accommodations. Whereas wilderness lovers such as Muir and Olmsted had dreamt of preserving an undefiled environment in which humankind could recapture its identity through direct contact with Nature, commercial exploitation has now made this impossible in some of the most spectacular areas of the national parks.

The white-water boating down the Colorado River

through the Grand Canyon of Arizona provides the most depressing example of prostitution of the wilderness experience. Travel agencies now offer river-running tours complete with shrimp and ice cream. Playboy "bunnies" are photographed running the river *au naturel.* Thus river running, which used to be a high-risk adventure through unknown country, has been reduced to a passive expedition through a Disneyland setting. The few real river runners who get a chance to cross the canyon on their own suffer not only from the loss of the wilderness experience but also from watching the physical degradation of the gorge from overuse.

The only solution to the overuse and degradation of wilderness areas is in restriction of visitors. The Forest Service and the National Park Service are already experimenting with rationing systems in certain areas. In the Grand Canyon the number of people authorized to take the river trip has been fixed. Backpackers must also obtain permits in order to minimize the impact on the fragile alpine type of ecosystem and to prevent overcrowding in the space available in the various backcountry grounds. Strict management and reduction of use is thus the price of popularity, but the quality of freedom associated with wilderness is diminished if not destroyed by human control of its use. We have reached a paradoxical situation, that we can save some of the wilderness experience only by introducing into wild areas the ordering and discipline that is becoming increasingly objectionable in civilized life.

Symbiosis of Humankind and the Earth

WHEREAS human management of Nature and awareness of environmental degradation are old stories, they began to attract wide attention only in our time. The reason for this long indifference on the part of scientists as well as the general public is that, until late in the nineteenth century, the effects of human intervention into Nature were limited in scope and intensity by the relative smallness of the world population and by the weakness of its means of action. It was still possible at the turn of the century to accept the ancient image of the universe in which, as Aristotle had maintained, everything had a definite place in an immutable hierarchy. A clear distinction between humankind and Nature was then taken for granted.

The popular image of the universe changed during the Renaissance and even more during the Enlightenment and the Industrial Revolution, when the feeling became prevalent that the human vocation was not only to manage Nature but to conquer and use it as a resource for the creation of wealth. Francis Bacon was the most articulate early exponent of this new demiurgic image, and he convinced most philosophers of the Enlightenment that knowledge was power and would be the chief contributor to the betterment of human life. The faith that knowledge would accelerate material progress led Benjamin Franklin to formulate as early as 1743 "a proposal for Promoting Useful Knowledge among the British Plantations in America." His plan was to establish an academy for discussions and experiments that, in his words, would "let Light into the Nature of Things, tend to increase the Power of Man over Matter, and multiply the Conveniences and Pleasures of Life." This concept resulted in the creation of the American Philosophical Society for the Promotion of Useful Knowledge, which obtained its charter in 1780.

Francis Bacon had early realized, however, that the increase in the "Power of Man over Matter" could be useful and safe only if humankind concerned itself with the long-range consequences of the practical applications of knowledge. In 1605, at the very beginning of the scientific era, he remarked in *The Advancement of Learning* that "the invention of the mariner's needle which gives the direction is of no less benefit for navigation than the invention of the sails which gives the motion." He believed in other words that the betterment of life would depend as much on the formulation of goals as on the development of techniques.

Admittedly, Bacon's warning has not had much influence on the promoters and technologists of the nineteenth and twentieth centuries. In general, modern people have shown greater interest in action than direction, but there is evidence that the social mood is beginning to change. While bigness and speed are still the most widely accepted criteria of success, we have come to realize that the word *progress* does not mean only moving forward. It may mean moving in the wrong direction.

The change in public attitude can be seen in the light of an event that occurred less than half a century ago. In 1933, the city of Chicago held a World's Fair to celebrate the "Century of Progress" that had elapsed since the city's birth and that had seen the triumph of scientific technology. The organizers of the fair were so convinced that scientific technology invariably improves human life that they stated in the guidebook, "Science discovers, genius invents, industry applies, and *man adapts himself to* or is *molded by* new things."

One of the subtitles of the guidebook was "Science Finds, Industry Applies, Man *Conforms*" [italics mine]. This philosophy was still dominant among futurologists of the 1950s when they tried to forecast what the world would be like in the year 2000. With dismal uniformity, they envisioned a future shaped by visionary technologies and architectures without relevance to human needs or natural conditions.

A fundamental change of attitude occurred during the 1960s and 1970s. No one would dare state today that humans must conform to technological imperatives or that they must accept being molded by technological forces We think instead that industrial development should be adapted to humankind and to Nature—not the other way around, as was advocated by the organizers of the Chicago Fair.

Charles Lindbergh's life, as recorded in his posthumously published *Autobiography of Values,* symbolizes how the modern world has begun to evolve from fascination with sophisticated technologies to the realization that unwise and excessive dependence on these technologies threatens fundamental values. While on a camping trip in Kenya during his late adult life, Lindbergh became intoxicated with the sensate qualities of African life that he perceived "in the dances of the Masai, in the prolifigacy of the Kikuyu, in the nakedness of boys and girls. You feel these qualities in the sun on your face and the dust on your feet . . . in the yelling of the hyenas and the bark of the zebras." Experiencing these qualities made Lindbergh ask himself, "Can it be that civilization is detrimental to human progress? . . . Does civilization eventually become such an overspecialized development of the intellect, so organized and artificial, so separated from the senses that it will be incapable of continued functioning?"

Lindbergh's doubts concerning civilization were the

more surprising to me because, in the 1930s I had known him as a colleague in the laboratories of the Rockefeller Institute for Medical Research, where he was developing an organ perfusion pump in collaboration with Dr. Alexis Carrel. His dominant interest at that time was, along with aviation, mechanical devices to explore what he calls in his book "the mechanics of life." His *Autobiography of Values* reveals how he eventually moved from an exclusive interest in the mechanical applications of science to concern for its social and philosophical implications. He remained enamored of modern science and was, for example, fascinated by space explorations, but he became increasingly distraught at seeing technology used for trivial and destructive ends.

Thus Bacon at the beginning of the scientific era and Lindbergh more than two centuries later expressed in different words a problem that has become central to our form of civilization. Science and technology provide us with the *means* to create almost anything we want, but the development of means without worthwhile *goals* generates at best a dreary life and may at worst lead to tragedy. Some of the most spectacular feats of scientific technology call to mind Captain Ahab's words in Melville's *Moby Dick:* "All my means are sane, my purpose and my goals mad." The demonic force, however, is not scientific technology itself, but our propensity to consider means as ends—an attitude symbolized by the fact that we measure success by the gross national product rather than by the quality of life and of the environment.

Many apparent achievements of our times are the current manifestations of trends initiated several decades ago. We have advanced civilization chiefly by accelerating and magnifying the process of change, but now we are faced with the absurdity of our so-called progress. One motorcar contributes to freedom; one hundred million motorcars not only generate traffic jams and pollution, but also become an addiction. Industrial energy in small doses makes life easier and more diversified; complete dependence on industrial energy amounts to a form of slavery. It is becoming more and more apparent that we can get our means and ends straight only by inquiring into the long-range consequences of our activities.

People concerned with such problems usually emphasize the damage done to the Earth by human activities, and to human life by pollution, overpopulation, or the depletion of resources. They regard the environment as foreign to human nature, with values of its own but separated from those of human life. With the growth of ecological sciences the image of man as the conqueror now tends to be replaced by the image of man as the seeker of a *modus vivendi* with the total physical and biological environment, the ultimate ideal being an intimate communion with the cosmos. In my opinion, however, this concept of *modus vivendi* does not convey the creativeness of the relationship that can exist between humankind and the Earth. Since my belief in this creativeness has its origin in the facts of biological evolution, I shall open here a large parenthesis to present, in somewhat technical terms, general remarks concerning the creative effects of living systems on the terrestrial environment.

As I have stated at greater length elsewhere, our planet owes its uniqueness in the solar system to the living things that it harbors. Surprising as it may sound, the terrestrial atmosphere and soil that are essential for the present forms of life were initially created and continue to be created by the various types of living things that have consecutively existed on Earth. If it were not for the continuous transformations of inanimate mineral and gaseous matter that existed before the appearance of life, the surface of the Earth would look as bleak and desolate as that of the Moon and Mars.

Organic substances similar to those present in all living things are widely distributed in the solar system, and perhaps even throughout the universe; moreover, they can be readily produced in the laboratory. But there is no knowledge, only unsubstantiated hypotheses, concerning the mechanisms through which lifeless matter gave rise to life. On the other hand, there is overwhelming evidence for a fundamental unity underlying all species that have lived in the past and are living now. It is even known that all the fundamental mechanisms for the biological production and use of energy, as well as for the synthesis and degradation of biostructures, emerged very early, more than 2 billion years ago. From then on, biological evolution seems to have proceeded not so much by creating biochemical novelties as by using and reassembling over and over again the same biochemical units, but according to different patterns. In the words of the French biologist François Jacob,

> What distinguishes a butterfly from a lion, a hen from a fly, or a worm from a whale is much less a difference in chemical constituents than in the organization and distribution of these constituents. The few big steps of evolution required acquisition of new information. But specialization and diversification occurred by using differently the same structural information.

The production of new living forms also occurred through another type of biological combination—symbiosis—which consists in the association of organisms having different genetic endowments. Etymologically, the word *symbiosis* simply means "living together," but when it was first introduced into the scientific literature a century ago, it denoted a more creative type of biological association, namely the integration of certain microscopic algae with microscopic fungi to produce larger and much more complex organisms called lichens. The microscopic algae and fungi retain their genetic identity while they are associated in the lichens, yet each particular lichen exhibits characteristics that transcend those of its two microbial constituents. Countless examples of such creative symbiotic partnerships exist at all levels of biological and social organization. Plants, for example, owe their ability to utilize solar energy to the presence in their cells of organelles called chloroplasts, which produce the chlorophyll responsible for photosynthesis. Genetically, chloroplasts are different from the plant cells in which they function. In the distant past, they were probably microorganisms capable of independent life, but they lost this capability during their long association with plants. The organelles called mitochondria that carry out indispensable metabolic functions also probably began as independent microorganisms before having become permanently associated with animal and plant cells. Like the microscopic algae and fungi in the lichens, chloroplasts and mitochondria retain their own genetic identity in the cells that harbor them, but their association with them is immensely creative and one of the main factors in the extension of life.

The most interesting aspects of living things are thus not their fundamental chemical components, which have their counterparts in the nonliving world, but their ability to undergo continuous modifications and associations. These attributes enable them to evolve into innumerable forms increasingly different from one another. It is in the course of biological evolution that all forms of life, even the simplest, have transformed the primeval lifeless environment from which they derived their sustenance. Life has created the present soils and atmosphere of the Earth out of the original rocks and gases of the initially lifeless planet. Reciprocally, the various forms of life have progressively undergone adaptive changes to fit the new conditions they have themselves created.

The manifestations of life in any organism thus imply the operations of fundamental ecological mechanisms and cannot be understood apart from the environmental factors to which the particular organism is adapted. As we shall now see, the biological evidence that the interplay between the various components of an ecological system can have creative effects applies to the reciprocal interplay between humankind and Earth.

The ecological image of human life that is now emerging is in part a consequence of concern for environmental degradation. It has also been influenced by the development of new scientific disciplines such as cybernetics, information sciences, general systems theory, and hierarchy theory. Its more profound origin, however, is the increased awareness of the intimate interdependence between human beings and their total environment. Starting from natural conditions, human societies have created cultural environments that in turn have influenced the course of their social evolution—a process of feedback characteristic of all ecological systems.

All over the world, the association between a given social group and a given environment has generated new social and environmental values. Both sides of the English Channel, for example, are similar in climate, topography, and geological structure. Their human populations also have essentially the same origin, being largely a mixture of Celtic, Scandinavian, and Mediterranean people. Yet England differs profoundly from France in the appearance of the countryside, management of agriculture, patterns of behavior, and traditional customs. These national characteristics are the consequences of historical choices that have determined unique types of relationships between people and their environment. The propensity of people on the French side of the Channel to cut long straight *allées* through the forest and to make wide use of pruning shears to shape fruit trees, grapevines, rosebushes, boxwood, and countless other plants certainly influences not only the physical appearance of the countryside but also the ways of thinking of its inhabitants. English people also control their physical and biological environment, but in a less geometric and Cartesian spirit. Hardly anything is known concerning this kind of environmental history.

The evolution of the Pacific Islands has created differences among them that are also the outcome of the interplay between the various human populations and the natural environments. Tahiti, Fiji, and Hawaii do not differ markedly from each other in climate or physical geography, but they have acquired distinctive characteristics that originate from the species of plants, types of agriculture, special technologies, and criteria of behavior that were introduced first by early native populations and later by French, English, or American colonists. In the Pacific islands, just as in the countries bordering the English Channel, the interplay between human attitudes and environmental conditions has generated new social values and new ecological systems that could not have emerged from natural forces.

Most natural situations can furthermore support simultaneously several distinct types of relationships between humankind and Earth, with the creation of as many different types of social and environmental values. For example, five different ethnic groups exist side by side on the semidesertic lands of the American Southwest: the Navajos, the Zunis, the Mormons, the Catholic Mexicans, and the largely Baptist Texas rangers. One could hardly imagine more contrast in ways of livelihood, social organizations, relations to nature, and religious systems than among these groups. Admittedly, there are ecological objections to many land use practices of these different peoples, but the point of interest is that while the ethnic groups function in similar natural environments under the same sky, they have created very different social and environmental values because they march to different cultural drums.

People have long been aware of the fascinating diversity of relationships between humankind and Earth. They have tried to account for it by myth, theology, philosophy, and some feeble attempts at scientific explanation. With awe and humility, pride or arrogance, they have regarded these variegated relationships as originating from a Creator who imposes certain patterns on life and matter, or as the inexorable expressions of evolutionary mechanisms. I find it more satisfactory to see humankind and Earth as constituting a diversity of systems of symbiosis that constantly undergo adaptive changes and thus contribute to a continuous evolutionary process of creation.

Noblesse Oblige

THE dangers of human intervention into nature have been repeatedly analyzed by modern scholars, most trenchantly perhaps by the American ecologist David Ehrenfeld in his recent book *The Arrogance of Humanism.* Ehrenfeld gives to the word *humanism* a broader meaning than the usual one. He uses it scornfully to denote the belief, almost universal in the countries of Western civilization, that we can engineer the future according to our whims, by arranging and re-arranging the natural world in any way we see fit. We operate on the belief that we can fully understand natural processes and therefore predict the consequences of our interventions into Nature. With numerous examples taken from contemporary life, Ehrenfeld has no difficulty in showing that this human conceit has generated dangerous situations that could not have been predicted, that probably cannot be corrected, and that may eventually destroy the most "humanistic" of modern societies.

Our arrogance, in Ehrenfeld's sense, is still very much at work—and not only in the United States. For example, plans are under discussion in the USSR to redirect toward the south the course of certain Siberian rivers that naturally flow north, so that their water can be used to irrigate vast areas of semidesertic land. Similarly, the president of a famous Japanese research institute has proposed under the title, "A Dream for Mankind," eight gigantic projects of environmental management that might go far, he thinks, toward solving the world's energy and food problems. One of these projects would be to dam the Congo River in central Africa so as to create an immense lake that would "improve" natural conditions. Another project would be to build a barrier across the Bering Strait between the United States and the USSR to interrupt the sea currents from the Arctic Ocean and thus make the Pacific climate more temperate. A third project would be to dam the Sampo River on the upper reaches of the Brahmaputra between China and the Indian province of Assam; the water would then be allowed to flow to India through a tunnel across the Himalayas to operate the largest hydroelectric power plant in the world.

It is easy to imagine the *local* benefits that may be derived from such enormous enterprises, but it is certainly impossible to predict their effects on the *global* ecosystem. This does not mean of course that they should not be undertaken. The Suez and Panama canals also constituted great ecological risks when they were built. If the modification of the surface of the Earth is a form of arrogance, this has been a feature of human life for immense periods of time. In fact, as mentioned earlier, people have no choice but to transform the surface of the Earth since they are not biologically adapted to most of the natural environments in which they live.

Ehrenfeld discusses some of the environmental values that are "being lost" as a result of present human activities. "British hedges and small fields, European farms and vineyards, North American urban and suburban parks and farms, and gardens everywhere are either being destroyed or altered in the name of efficiency." All the ecosystems mentioned in this passage are, in his words, "cultured landscapes," and some of them are of recent origin. Objecting to the destruction or alteration of these artificial ecosystems merely because of opposition to *any* change is an expression of the attitude that I labeled "environmental ambivalence" at the beginning of this chapter, or more simply of the desire to keep everything as it was during one's youth. Change will continue to be because it is an inevitable condition of life. But it must be based on good reasons. I believe with Ehrenfeld that destroying or altering existing ecosystems, whether natural or artificial, in the name of "efficiency" can often have unfortunate consequences.

Since manipulating Nature is an inevitable aspect of the human condition, it is a natural attitude and not a manifestation of arrogance, especially when efforts are made to base action on knowledge and judgment of values. I have been allowed by Rufus E. Miles, Senior Fellow of the Woodrow Wilson School of Public and International Affairs at Princeton University, to quote from a letter he wrote me after participating in a conference during which Ehrenfeld and I expressed slightly different views concerning human interventions into nature:

> I would much have preferred to have you mention the concept of "noblesse oblige" as man's appropriate role at the top of the animal kingdom. A person of noble birth and outlook learns to accept his elevated status and knows that others will serve him, yet he treats them with thoughtfulness and kindness. He accepts a reciprocal responsibility toward them. This is a far cry from arrogance. Do you not think it would be well to preserve the word "arrogance" for its intended usage, which is a vain, condescending, and unkind form of behavior by a person in a position of power toward other persons or fauna or flora?

Miles's use of the expression "noblesse oblige" seems to me an admirable way of expressing the attitude with which we should approach all environmental problems. We shall continue to intervene into nature, but we must do it with a sense of responsibility for the welfare of the Earth as well as of humankind, and we must therefore attempt to anticipate the long-range consequences of our actions. Human modifications of the Earth can be lastingly successful only if their effects are adapted to the invariants of physical and human nature. Fortunately, such constraints are compatible with diversity; there are many ways to deal with nature that accord with natural laws. A forest in the temperate region lends itself to the creation of parks as different in style as those of England, France, and Japan. In England, the so-called New Forest has been under constant management since 1079, and different parts of it are treated according to different ecological formulae—some left *au naturel,* some carefully pruned, some reserved for recreational activities, and so on. The individuality of a cultural environment is achieved through the choices made by a particular culture among the several options available to it at a given time in a given place.

Until recently, options were provided almost exclusively by the natural world and choices were made by caprice or empirical wisdom. Increasingly now, options and choices are affected by scientific knowledge. In many countries of Western civilization, for example, the higher yields of scientific agriculture have led to the abandonment of marginal farmland. It may soon be economically possible to use this abandoned land for the production of rapidly growing trees or other vegetation that can serve as sources of energy, and of raw material for chemical industries. On the other hand, it will be more desirable in other places to keep the land clear of brush so as to use it for recreational purposes or for aesthetic pleasure. Depending upon circumstances, this can be done by keeping animals grazing on it, by annual mowing, or by treatment with chemicals.

Knowledge enlarges the range of options through different mechanisms. It enters the public domain in the form of verifiable facts and laws. It generates innovations that can help to achieve chosen ends. It constantly surprises and subverts because new discoveries and applications are largely unpredictable. Knowledge thus makes people more receptive to new attitudes and more willing to change their ways. Who would have predicted two centuries ago that the soybean and the potato would come to occupy such a large place in the economy of Western nations, or that population control would first become possible on a large scale through the use of a contraceptive prepared from a seemingly trivial plant growing wild in Mexico!

However, while knowledge increases the range of options, it cannot be the sole basis for decision making, because it is always incomplete and therefore cannot describe all aspects of the world that bear on human life and environmental quality. Knowledge is more effective as a generator of possibilities than as a guide to choice and as a source of ethics. In the final analysis, the management of the Earth must be value conscious and value oriented. Human systems are different from natural systems in that they are teleological as well as ecological. Each human society has its own images of the future that influence its policies. The extent of our interventions into nature is inevitably influenced, for example, by social attitudes concerning natural resources.

Although technological civilization depends upon abundant supplies of metals and energy, opinions differ as to the priority connected with obtaining these supplies in a wider context. Large reserves of copper exist in Cascades National

Park, but their exploitation would require a huge open mine that would spoil a wonderful wilderness area. Titanium could be obtained from the sand of Cape Cod, and various other metals as well as uranium from the granite of the White Mountains, but at the expense of the aesthetic appeal of these humanized landscapes. In these cases, then, the "limits to growth" are determined not by availability of raw materials but by the choices society makes concerning other factors affecting the quality of life.

The present supplies of fossil fuels will eventually be depleted, but practical techniques will certainly be developed within a few decades to produce energy from renewable sources—nuclear or solar or perhaps both. The selection of methods for the production and use of energy will involve choices, however, based not only on scientific knowledge and cost-benefit analysis, but also on judgments of value. For example, the development of nuclear energy will inevitably require enormous generators that will lead to strict technological and social controls. Similarly, trapping solar energy in orbiting satellites and beaming it on Earth in the form of microwaves that will be used to generate electricity will require a high degree of social organization and centralization. In contrast, the first steps in the use of conventional sources of solar energy will have to be carried out in fairly small units—a necessity that will favor social decentralization. Many people, perhaps the great majority, will prefer abundant electricity without giving thought to its origin, its environmental effects, and its indirect social costs. Other people will prefer instead small-scale technologies compatible with social decentralization and regional and cultural pluralism. The final outcome will probably be a mix of centralized and decentralized sources of energy, selected to fit the environmental and social characteristics of each given situation, and facilitating the expression of the multiple aspects of human nature.

The production and use of resources and energy are not the only factors that will affect management of the Earth. Whatever types of intervention into Nature are being contemplated, the main concerns of their proponents have so far been productivity and efficiency because these are the measures of technical perfection and economic reward. Yet the really significant end products of the changes resulting from human interventions are their long-range effects on the quality of human life and the environment. Efficiency and productivity have been identified with progress because they have contributed to wealth but we now realize that they are often achieved at great human and environmental cost. The squalor of Coketown in Dickens's *Hard Times;* the misery of the people, the land, and the streams in Appalachia, and all the other horror stories of the present ecological crisis make it clear that the words *productivity* and *efficiency* can serve as a measure of real success only if they incorporate, along with their purely technical connotations, concern for biological and social applications and values.

From the point of view of human and environmental quality, it is probable that diversity and flexibility are more valuable than productivity and efficiency. Natural ecosystems are characterized not by high productivity but by resilience and flexibility, attributes that enable them to persist in the face of climatic disturbances and other uncertainties. Ecosystems are more likely to be durable when they contain a great diversity of species and when these are linked in complex symbiotic relationships. Furthermore, biological progress through evolutionary mechanisms tends to be more rapid when the total population is subdivided into colonies that are sufficiently small and separated to permit the survival of mutant forms, yet sufficiently interconnected to permit interbreeding.

Human societies differ from natural ecosystems in that they are influenced by teleological considerations at least as much as by environmental conditions. There are, nevertheless, suggestive analogies between human and natural systems. Just as biological diversity facilitates Darwinian evolution, so is cultural diversity essential for social progress. It is probably fortunate in this regard that, all over the world, ethnic and regional groups are asserting their identity and are beginning to recapture some autonomy. This might help increase cultural diversity and thereby the rate of creative social change.

Even if the trend toward social decentralization is successful, however, there will probably occur simultaneously an increasing globalization of certain types of human activities, in particular those dealing with transportation and other forms of communication, and those dependent on large-scale complex technologies. We may thus gradually move toward a dual type of human relationship between humankind and Earth—on the one hand an increasingly centralized management based on the use of highly automated technologies derived from sophisticated science, and on the other hand a decentralized management dealing at the local level and on the human scale with the more intimate aspects of life. The general formula of management for the future might be, think globally and act locally.

Global thinking and local action both require understanding of ecological systems, but ecological management can be effective only if it takes into consideration the visceral and spiritual values that link us to the Earth. Scientifically defined, ecology is nothing more than the study of interrelationships between living things and their environment; it is therefore ethically neutral. These relationships, however, are always influenced by the human presence, which introduces an ethical component into all environmental problems. Since the nature of our activities determines the extent and direction of environmental changes, ecological thinking must be supplemented by humanistic value judgments concerning the effect of our choices and actions on the quality of the relationship between humankind and Earth, in the future as well as in the present. Noblesse oblige.

CHAPTER NINE

Envoi

DURING the Old Stone Age, the human species lived in small bands that obtained their subsistence from hunting, fishing, and collecting wild plants. For the past ten thousand years, the immense majority of human beings have practiced agriculture, pastoralism, and a variety of small trades, in villages or nomadic groups small enough to allow close social relationships and direct contact with nature. In recent times, the urban environment has given them new comforts, larger social contacts, and greater scope for the expression of individuality.

These different ways of life have left their stamp on human nature, in part through genetic coding but chiefly through physiological and social conditioning. As a consequence of this complex history of our species, most human beings long to recapture now and then each of the various experiences of their evolutionary past: that of the hunter-gatherer, of the farmer and pastoralist, and of the urban dweller. The wooing of the Earth thus implies much more than converting the wilderness into humanized environments. It means also preserving natural environments in which to experience mysteries transcending daily life and from which to recapture, in a Proustian kind of remembrance, the awareness of the cosmic forces that have shaped humankind.

We cannot escape from the past, but neither can we avoid inventing the future. With our knowledge and a sense of responsibility for the welfare of humankind and the Earth, we can create new environments that are ecologically sound, aesthetically satisfying, economically rewarding, and favorable to the continued growth of civilization. But the wooing of the Earth will have a lastingly successful outcome only if we create conditions in which both humankind and the Earth retain the essence of their wildness. The symbiosis between these two different but complementary expressions of wildness will constantly engender unexpected values and new hopes, in an endless process of evolutionary creation.

Can the Earth Be Saved?

APPENDIX I

As indicated in Chapter 3, page 30, I am listing here by rank of importance according to my judgment some of the dangers that threaten humankind and the Earth. I have used two criteria for evaluating the relative importance of these dangers: on the one hand the extent of expected damage from them, and on the other, the extent of our ignorance concerning their control.*

Nuclear warfare

I do not see that any significant progress has been made toward its prevention.

Failure to provide meaningful employment for young people

This tragic situation goes far deeper than lack of jobs; its ultimate consequence is desocialization and dehumanization.

Overpopulation

Universal shortages of food, resources, and livable space will inevitably occur and generate global conflicts if the world population continues to increase at the present rate. There is reason to hope, however, that the population will stabilize and may even decrease in many places before it reaches catastrophic levels.

*I am aware of the possibility of disastrous climatic changes, both of natural and human origin. But I have not listed this danger because there is not sufficient knowledge to evaluate the likelihood of its occurrence.

APPENDIX I

Environmental degradation

If carried much further, the destruction of the tropical rainforest will profoundly disturb the global ecosystem and impoverish the genetic diversity of the biota. Furthermore, the agricultural potentialities of the Earth are being rapidly decreased by desertification and other forms of damage to farmland. Awareness of these problems has led to remedial action in some places.

Excessive use of energy and resources

Although shortages of energy and resources will certainly occur, it is probable that new technologies will be developed to produce what we really need. The greatest dangers for the future, in my opinion, will come not from shortages of energy and resources but from their overuse and misuse.

Environmental pollution

Although global pollution is still increasing, there is hope that it will not reach catastrophic levels, because awareness of its occurrence and consequences is widespread and is beginning to be reflected in better technological practices. Furthermore, methods of pollution control can be developed and applied within present economic systems.

*Selected Successes and Associated Problems of the Technological/Industrial Era**

APPENDIX II

"Successes"	Problems Resulting from Being "Too Successful"
Prolonging the lifespan, reducing infant mortality	Regional overpopulation, problems of the aged
Highly developed science and technology	Hazard of mass destruction through nuclear and biological weapons, threats to privacy and freedoms (e.g., surveillance technology, "bioengineering")
Advances in communication and transportation	Increasing air, noise, and land pollution, "information overload," vulnerability of a complex society to breakdown
Efficient production systems	Dehumanization of ordinary work
Affluence, material growth	Increased per capita consumption of energy and goods, leading to pollution and depletion of the Earth's resources
Satisfaction of basic needs	Worldwide revolutions of "rising expectations," rebellion against nonmeaningful work
Expanded power of human choice	Unanticipated consequences of technological applications, management breakdown as regards control of these
Expanded wealth of developed nations	Increasing the gap between "have" and "have-not" nations, frustration of the "revolution of rising expectations"

*Adapted from Willis W. Harman, "On Nominative Forecasting," in Wayne I. Boucher, ed., *The Study of the Future: An Agenda for Research.* Washington, D.C.: NSF/RA770036, 1977, p. 80.

APPENDIX III

For the reasons given on pages 70–72, I do not believe that the biblical doctrine of man's dominion over nature has played a significant role—if any at all—in the exploitation of the Earth's resources or in the degradation of the environment. I am listing here some of the books and articles I have read on this topic in the hope that they will stimulate a more rigorous analysis of the influence that religious beliefs or doctrines have exerted on human attitudes toward the Earth.

Anon. *Man and Nature.* A report submitted to the Archbishop of Canterbury on the relevance of Christian doctrine to the problems of man in his environment. London: Collins, 1975.

Bailey, L. H. *This Holy Earth.* New York: Scribners, 1915.

Barbour, Ian, ed. *Western Man and Environmental Ethics.* Reading, Mass.: Addison-Wesley, 1973.

Bennett, John. *Cultural Anthropology and Human Adaptation.* New York: Pergamon Press, 1976.

Black, John. *The Dominion of Man: The Search for Ecological Responsibility.* Edinburgh: Edinburgh University Press, 1970.

Bonifazi, Conrad. "A Theology of Things," in Robert H. McCabe and R. F. Mines, eds. *Man and Environment,* vol. 2. Englewood Cliffs, N. J.: Prentice-Hall, 1974, p. 368.

———. "Biblical Roots of an Ecological Conscience," in Michael Hamilton, ed. *This Little Planet.* New York: Scribners, 1970, p. 203.

Burghardt, W. J. *Towards Reconciliation.* Washington, D. C.: United States Catholic Conference, 1974.

Burstein, Samuel M. "Science, Abraham, and Ecology." *Bulletin of the Atomic Scientists* 28 (1972), 88.

Dillistone, F. W. *Charles Raven.* Grand Rapids, Mich.: W. B. Eerdman, 1975.

Dubos, René. "Franciscan Conservation Versus Benedictine Stewardship," in *A God Within,* Chapter 8. New York: Scribners, 1972.

APPENDIX III

Elder, Frederick. *Crisis in Eden.* Quoted in McCabe and Mines, vol. 2, p. 363.

Foley, Michael. "Who cut down the sacred tree?" *The CoEvolution Quarterly,* Fall 1977, p. 62.

Follansbee, Lewis. "The antecedents of contemporary problems and solutions in ecology," in McCabe and Mines, vol. 2, p. 381.

Friedman, Maurice. "Social Responsibility in Judaism." *Journal of Religion and Health* 2 (1962).

Fritsch, Albert J. *A Theology of the Earth.* Washington, D.C.: CLB Publishers, 1972.

Glacken, Clarence J. "This Growing Second World Within the World of Nature," in F. R. Fosberg, ed. *Man's Place in the Island Ecosystem.* Honolulu: Bishop Museum Press, 1965, p. 75.

———. "Man's Place in Nature in Recent Western Thought," in Hamilton, p. 163.

———. *Traces on the Rhodian Shore.* Berkeley: University of California Press, 1967.

Gordis, Robert. "The Earth is the Lord's—Judaism and the Spoliation of Nature." *Keeping Posted* 16 (1970), 5.

Hamilton, Michael, ed. *This Little Planet.* New York: Scribners, 1970.

Hatch, R. McC. "Cornerstones to a Conservation Ethic." *Atlantic Naturalist* 12(1957), 154.

Hughes, J. Donald. *Ecology in Ancient Civilizations.* Albuquerque: University of New Mexico Press, 1975.

Lowe, George. "Belief Systems," in McCabe and Mines, vol. 2, p. 359.

Marney, Carlyle. *A Christian Life-Style?* Chapel Hill, N.C.: Duke University, 1977.

McCabe, Robert H. and R. F. Mines, eds. *Man and Environment,* vol. 2. Englewood Cliffs, N. J.: Prentice-Hall, 1974.

Moore, R. E. *Man in the Environment.* New York: Knopf, 1975.

Nasr, Seyyed Hossein. *The Encounter of Man and Nature.* London: Allen and Unwin, 1968.

Needleman, Jacob. *A Sense of the Cosmos: The Encounter of Modern Science and Ancient Truth.* Garden City, N.Y.: Doubleday, 1975.

O'Connor, Daniel and Francis Oakley. *Creation: The Impact of an Idea.* New York: Scribners, 1969.

Passmore, John. *Man's Responsibility for Nature.* New York: Scribners, 1974.

———. *The Perfectibility of Man.* New York: Scribners, 1970.
Raven, Charles. *Natural Religion and Christian Theology.* Cambridge: Cambridge University Press, 1953.
———. *Teilhard de Chardin—Scientist and Seer.* London: Collins, 1962.
Rose, David J. "The World Won't Stop and We Can't Jump Off." *The Episcopalian,* June 1979, p. 6.
Santmire, Paul H. *Brother Earth.* New York: T. Nelson, 1970.
Schumacher, E. F. *Small is Beautiful.* New York: Harper & Row, 1973.
Sessions, George. "Anthropocentrism and the Environmental Crisis." *Humboldt Journal of Social Relations* 2 (1974), 1.
Shepherd, J. B. "Theology for Ecology," *Catholic World* 210 (1970), 173.
Shinn, Roger. "The Wind and the Whirlwind," in Ian G. Barbour, ed. *Finite Resources and the Human Future.* Minneapolis: Augsburg, 1976
Sittler, Joseph. "Ecological Commitment as Theological Responsibility,' *Idoc* (1970), 76.
Sopher, David E. *Geography of Religions.* Englewood Cliffs, N. J.: Prentice-Hall, 1967.
Spring, David and Eileen Spring. *Ecology and Religion in History.* New York: Harper & Row, 1974.
Suzuki, D. T. "The Role of Nature in Zen Buddhism." *Eranos-Jahrbuch* 22 (1953), 291.
Thompson, William Irwin. *Passages About Earth.* New York: Harper & Row, 1973.
Tuan, Yi-Fu. "Our Treatment of the Environment in Ideal and Actuality." *American Scientist* 58 (1970), 244.
Vaux, Kenneth. *Subduing the Cosmos: Cybernetics and Man's Future.* Richmond, Va.: John Knox Press, 1970.
Vorspan, Albert. *The Crisis of Ecology: Judaism and the Environment.* New York: Union of American Hebrew Congregations, 1970.
Warner, Langdon. "Craftsman, Nature and Shinto," in *Japanese Sculpture of the Tempyo Period.* Cambridge: Harvard University Press, 1959.
White, Lynn. "The Historical Roots of our Ecologic Crisis," *Science* 155 (1967), 1203.

Since the compiling of this Appendix, a bibliography on this same topic has been published: "Questions of Christianity and Technology: A Bibliographic Introduction," *Science, Technology, and Society.* Lehigh University, 14 Nov. 1979.

NOTES

Epigraph

Rabindranath Tagore, *Towards Universal Man* (New York: Asia Publishing House, 1961), p. 294.

André Malraux, *Les Voix du Silence* (Paris: La Galerie de la Pléiade, 1951), p. 638.

Preface

Page XV, line 4

Robert Frost, "The Aim was Song," in Edward Lathem, *The Poetry of Robert Frost* (New York: Holt, Rinehart, and Winston, 1969), p. 223.

CHAPTER ONE. *A Family of Landscapes*

Page 3, line 12:

Plato, *Critias,* translated by B. Jowett (Oxford: Clarendon Press, 1953), vol. 3, p. 794.

Page 4, line 3:

Plato, *Phaedrus,* translated by B. Jowett (Oxford: Clarendon Press, 1953), vol. 3, p. 135.

Page 4, line 12:

Henry Miller, *The Colossus of Maroussi* (New York: New Directions, 1958), pp. 45, 48.

Page 4, line 20:

Costis (Kostes) Palamas, "Le Satyr ou la chanson nue," *Choix des Poésies* (Paris: Athènes, 1930); alsó in "Permanence de la Grèce," *Cahiers du Sud,* 1948, p. 288.

NOTES

Page 6, line 1:
Edward Steichen, *The Family of Man* (New York: Museum of Modern Art, 1955).

CHAPTER TWO. *The Wilderness Experience*

Page 9, line 31:
José Ortega y Gasset, *Meditations on Hunting,* translated by Howard B. Wescott (New York: Scribners, 1972), p. 142.

Page 10, line 8:
B. H. Lopez, *Of Wolves and Men* (New York: Scribners, 1978).

Page 11, line 14:
Puritan views of the forest quoted in E. T. Carlson and P. C. Novel, "Stress and Behavior in the Founding Pilgrims," *Bulletin of the New York Academy of Medicine* 47 (1971), 149. See also Marjorie Hope Nicolson, *Mountain Gloom and Mountain Glory* (Ithaca, N. Y.: Cornell University Press, 1959).

Page 13, line 2:
Conrad Gesner, quoted in Thomas F. Hornbein, *Everest: The West Ridge,* David Brower, ed. (New York: Ballantine Books, 1968), p. 70.

Page 14, line 1:
Francesco Petrarca, "The Ascent of Mt. Ventoux," in Ernst Cassirer et al., eds., *The Renaissance Philosophy of Man* (Chicago: University of Chicago Press, 1956), p. 36.

Page 14, line 13:
Charles Cotton, quoted in J. Passmore, *Man's Responsibility for Nature* (New York: Scribners, 1974), p. 107.

Page 15, line 2:
Henry Thoreau, "Walking," *Atlantic Monthly,* June 1862.

NOTES

Page 15, line 9:
Henry Thoreau, *The Maine Woods* (New York: Crowell, 1961).

Page 15, line 24:
Aldous Huxley, "Wordsworth and the Tropics," in *Do What You Will* (New York: Doubleday, 1929).

Page 16, line 13:
David Ehrenfeld, "Man's Intervention in Living Systems," in Ruth Kreplick et al., eds., *Proceedings of Engineering for the Environment,* Wakefield, Mass.: May 6–7, 1971; Helmuth Lieth and Robert H. Whittaker, eds., *Primary Productivity of the Biosphere* (New York: Springer-Verlag, 1975); Edmund A. Schofield, "Natural Systems," in E. A. Schofield, ed., *Earthcare: Global Protection of Natural Areas* (Boulder, Colo.: Westview Press, 1979), p. 75.

Page 17, line 24:
René Dubos, "Chapter 8: Franciscan Conservation vs. Benedictine Stewardship," *A God Within* (New York: Scribners, 1972).

CHAPTER THREE. *Can the World Be Saved?*

Page 19, line 15:
Senio C. Scott, *Fishing in American Waters,* 1875, quoted by Robert Cushman Murphy in *The Congressional Record,* Remarks of the Hon. Otis Pike, October 6, 1970.

Page 20, line 4:
Charles Dickens, *Hard Times* (New York: Norton, 1966), see for example pp. 17 or 48.

Page 20, line 19:
Thomas B. Johansson, "Atmospheric Pollution and Energy Production," in E. A. Schofield, ed., *Earthcare: Global Protection of Natural Areas* (Boulder, Colo.: Westview Press, 1979), p. 105.

NOTES

Page 22, line 15:
Robert Cushman Murphy, quoted by the Hon. Otis Pike in *The Congressional Record,* October 6, 1970.

Page 22, last line:
C. M. Vadrot, *Mort de la Méditerranée* (Paris: Seuil, 1978).

Page 23, line 2:
Lester R. Brown, *The Twenty-Ninth Day: Accommodating Human Needs and Numbers to the Earth's Resources* (New York: Norton, 1978); and "The Worldwide Loss of Cropland," Worldwatch Paper 24 (Washington, D.C.: Worldwatch Institute, October 1978); Erik Eckholm, *Losing Ground: Environmental Stress and World Food Prospects* (New York: Norton, 1976).

Page 23, line 6:
Kai Curry-Lindahl, "Conservation Problems of Savannahs and Other Grasslands," in Schofield, ed., *Earthcare,* 359; Michael H. Glantz, ed., *Desertification: Environmental Degradation in and around Arid Lands* (Boulder, Colo.: Westview Press, 1977); Henri-Noël Le Houérou, "La désertisation des régions arides," *La Recherche* 10 (1979), 336; Anders Rapp, *Desertification in Africa,* reviewed in *The International Journal of Environmental Studies* 9 (1976), 75.

Page 25, line 16:
Lee Merriam Talbot, "Wilderness Overseas," *Sierra Club Bulletin* 42 (1957), 28.

Page 26, line 8:
R. J. A. Goodland and H. S. Irwin, *Amazon Jungle: Green Hell to Red Desert* (New York: Elsevier, 1975); Daniel H. Janzen, "Whither Tropical Ecology," in John A. Behnke, ed., *Challenging Biological Problems* (New York: Oxford University Press, 1972), p. 281; T. C. Whitmore, *Tropical Rain Forests of the Far East* (London: Oxford University Press, 1975).

Page 26, line 16:
Erik Eckholm, "Disappearing Species: The Social Challenge," Worldwatch Paper 22 (Washington, D.C.: Worldwatch Institute, July 1978);

Norman Myers, *The Sinking Ark: A New Look at the Problem of Disappearing Species* (New York: Pergamon Press, 1979).

Page 28, line 24:
Charles S. Aiken, "Faulkner's Yoknapatawpha County: Geographical Fact into Fiction," *The Geographical Review* 67 (1977), 1.

CHAPTER FOUR. *The Resilience of Nature*

Page 32, line 7:
Aldo Leopold, *A Sand County Almanac* (New York: Oxford University Press, 1949), p. 189.

Page 32, line 16:
Norman Myers, *The Sinking Ark: A New Look at the Problem of Disappearing Species* (New York: Pergamon Press, 1979).

Page 32, line 20:
J. Cairns, Jr., K. L. Dickson, and E. E. Herricks, eds., *Recovery and Restoration of Damaged Ecosystems* (Charlottesville: University Press of Virginia, 1977); Edward H. Graham, "The Recreative Power of Plant Communities," in W. Thomas, ed., *Man's Role in Changing the Face of the Earth* (Chicago: University of Chicago Press, 1956), p. 689; C. S. Holling, "Resilience and Stability of Ecological Systems," *Annual Review of Ecology and Systematics* 4 (1973), 1.

Page 33, line 23:
John J. Kupa and William R. Whitman, *Land-cover Types of Rhode Island: An Ecological Inventory* (Kingston: University of Rhode Island Agricultural Experiment Station Bulletin 409, 1972).

Page 35, line 13:
Lee W. DeCraff, "Return of the Wild Turkey," *The Conservationist* (October-November 1973), p. 24; Gerald A. Wuns and Arnold H. Hayden, "Turkey Renaissance," *Natural History* 82 (1973), 86.

NOTES

Page 35, line 27:
Lars Emmelin, "The Beaver-Conservation Problems," *Current Sweden*, Environment Planning and Conservation 61, February 1976. Available from Swedish Information Service, New York.

Page 36, line 14:
Erik P. Eckholm, *Losing Ground: Environmental Stress and World Food Prospects* (New York: Norton, 1976); Erik P. Eckholm, "The Deterioration of Mountain Environments," *Science* 189 (1975), 764; Edward G. Farnworth and Frank B. Golley, eds., *Fragile Ecosystems: Evaluation of Research and Applications in the Neotropics* (New York: Springer-Verlag, 1974); David Pimentel et al., "Land Degradation: Effects on Food and Energy Resources," *Science* 194 (1976), 149.

Page 37, line 25:
Sturla Fridriksson, *Surtsey: Evolution of Life on a Volcanic Island* (London: Butterworth, 1975).

Page 38, line 4:
Samir I. Ghabbour, "National Parks in Arab Countries," *Environmental Conservation* 2 (1975), 45.

Page 40, line 4:
Private communication.

Page 40, line 24:
Mohamed El-Kassas, *The International Pahlavi Environment Prize, 1978* (New York: United Nations Environment Programme, 1978), p. 30.

Page 41, line 3:
Cairns, Dickson, and Herricks, *Recovery and Restoration*, pp. 24–134; Trevor Halloway, "Rescuing our Rivers: Back from the Dead," *Environment* 20 (1978), 6; Thomas H. Maugh, "Restoring Damaged Lakes," *Science* 203 (1979), 425.

Page 44, line 15:
D. F. Costello, *The Prairie World* (New York: Crowell, 1969); David J.

Parsons, "The Role of Fire in Natural Communities: An Example from the Southern Sierra Nevada, California," *Environmental Conservation* 3 (1976), 91; H. E. Wright, Jr., "Landscape Development, Forest Fires, and Wilderness Management," *Science* 186 (1974), 487.

Page 45, line 13:
David Ehrenfeld, "Man's Intervention in Living Systems," in Ruth Kreplick et al., eds., *Proceedings of Engineering for the Environment,* Wakefield, Mass.; May 6–7, 1971.

Page 46, line 27:
Jon Tinker, "Farming and Conservation," *New Scientist* 62 (1974), 219.

Page 47, line 3:
Nicholas Mirov, "The Pines of Ravenna," *Natural History* 80 (1971), 24.

Page 48, line 3:
Roy Simonson, "The Soil Under Natural and Cultural Environments," *Journal of Soil and Water Conservation* 6 (1951).

CHAPTER FIVE. *Humanization of the Earth*

Page 49, line 13:
Rabindranath Tagore, *Towards Universal Man* (New York: Asia Publishing House, 1961), p. 294.

Page 55, line 6:
David Costello, *The Prairie World* (New York: Crowell, 1969).

Page 56, line 7:
Paul Brooks, *The View from Lincoln Hill: Man and the Land in a New England Town* (Boston: Houghton Mifflin, 1976).

Page 56, line 23:
Pierre Dansereau, "Ecological Impact and Human Ecology," in F. Fraser

Darling and John P. Milton, eds., *Future Environments of North America* (New York: The Natural History Press, 1966), p. 423.

Page 59, line 14:
Jay Appleton, *The Experience of Landscape* (New York: Wiley, 1975).

Page 65, line 22:
John Passmore, *Man's Responsibility for Nature* (New York: Scribners, 1974); Yi-Fu Tuan, "Our Treatment of the Environment in Ideal and Actuality," *American Scientist* 58 (1970), 244.

Page 67, line 7:
Peter Farb, "Rise and Fall of the Indians of the Wild West," *Natural History* 82 (1968), 34; Calvin Martin, *Keepers of the Game: Indian-Animal Relationships and the Fur Trade* (Berkeley: University of California Press, 1978).

Page 67, line 21:
C. W. Nicol, *From the Roof of Africa* (New York: Knopf, 1971).

Page 70, line 27:
D. T. Suzuki, "The Role of Nature in Zen Buddhism," *Eranos-Jahrbuch* 22 (1953), 291.

Page 70, line 29:
Lynn White, Jr., "The Historical Roots of our Ecologic Crisis," *Science* 155 (1967), 1203.

Page 73, line 22:
George Perkins Marsh, *The Earth as Modified by Human Action: A New Edition of "Man and Nature"* (New York: Scribner, Armstrong & Co., 1874).

Page 74, line 16:
Paul Chabrol, "L'Homme et son sens de la nature (de Jean Jacques Rousseau à nos jours)," *Académie des Sciences de Toulouse*, 23 February 1978, 44.

NOTES

CHAPTER SIX. *The Management of Earth*

Page 82, line 20:
Don Goldman, "A critique of 'Humanizing the Earth,' " *American Foresters* 79 (1973), 43.

Page 82, line 20:
René Dubos, "Humanizing the Earth," *Science* 179 (1973), 769.

Page 84, line 19:
Aldo Leopold, *A Sand County Almanac* (New York: Oxford University Press, 1949). p. 224.

Page 85, line 27:
Leopold, *A Sand County Almanac,* p. 219.

Page 86, line 22:
F. R. Fosberg, ed., *Man's Place in the Island Ecosystem* (Honolulu: Bishop Museum Press, 1965).

Page 87, line 3:
David Ehrenfeld, *Conserving Life on Earth* (New York: Oxford University Press, 1972); Edmund A. Schofield, ed., *Earthcare: Global Protection of Natural Areas* (Boulder, Colo.: Westview Press, 1979).

Page 89, line 30:
M. Evenari, L. Shanan, and N. Tadmor, *The Negev: The Challenge of a Desert* (Cambridge: Harvard University Press, 1971); Avraham Harman, "Agricultural Development," *Israel Today* No. 2, Jerusalem, April 1970; Efraim Orni and Dan H. Yaalon, "Reclamation and Conservation of the Soil," *Israel Today* No. 26, Jerusalem, December 1970.

Page 90, line 2:
M. Iman, "Mareotis: A Productive Coastal Desert in Egypt," in Schofield, *Earthcare,* p. 451.

NOTES

Page 91, line 3:
F. W. Went, "Plant Life and Desertification," *International Journal of Environmental Conservation,* Winter 1978, p. 263.

Page 97, line 10:
Donn Small, "Conservation and Recreation in the New Forest," *Journal of the Royal Society of Arts,* April 1978, p. 279.

Page 97, line 17:
Erik Eckholm, "Planting for the Future: Forestry for Human Needs," World Watch Paper 26 (Washington, D.C.: Worldwatch Institute, February 1979); Gunnar Poulsen, "A Leafy Paradox for Science," *IRDC Reports,* Ottawa, 7 (1978), 18.

Page 99, line 11:
Lawrence Halprin, "Israel: The Man-Made Landscape," *Landscape* 9 (1959–1960), 19, 23.

Page 100, line 29:
Major W. A. Jones, "Plan for the Cultivation of Trees Upon the Presidio Reservation," Engineer Office, Hq. Dept. of Calif., PSF—to Asst. Adj. Gen., Hq. Dept. of California, PSFC, 26 March 1883.

Page 101, line 28:
Geoffrey and Susan Jellicoe, *The Landscape of Man: Shaping the Environment from Prehistory to the Present Day* (New York: Viking, 1975).

Page 103, line 18:
John Ormsbee Simmonds, *Earthscape: A Manual of Environmental Planning* (New York: McGraw-Hill, 1978).

Page 104, line 7:
Richard Mabey, *The Unofficial Countryside* (London: Collins, 1973).

Page 104, line 16:
Elizabeth Barlow, *Frederick Law Olmsted's New York* (New York: Praeger, 1972).

Page 105, line 12:
Kenneth Mellanby, "Rus in urbe," *Nature* 278 (1979), 8.

CHAPTER SEVEN. *Of Places, Parks, and Human Nature*

Page 111, line 28:
Quoted in Arthur A. Ekirch, Jr., *Man and Nature in America* (New York: Columbia University Press, 1963), p. 12.

Page 112, last line:
Richard Wilbur, quoted in Alan Gussow, *A Sense of Place: The Artist and the American Land* (San Francisco: Friends of the Earth, 1971).

Page 117, line 29:
Bernard Rudofsky, *Architecture without Architects* (New York: Museum of Modern Art, 1964); and *The Prodigious Builders* (New York: Harcourt, Brace, Jovanovich, 1977).

Page 118, line 18:
C. A. Doxiadis, *Ecology and Ekistics* (Boulder, Colo.: Westview Press, 1976).

Page 119, line 26:
John Ruskin, *Modern Painters,* in E. T. Cook and Alexander Wedderburn, eds., *The Works of Ruskin* (London: George Allen, 1903–1912), vol. 5, p. 234.

Page 125, line 17:
Christopher Tunnard, *A World with a View: An Inquiry into the Nature of Scenic Values* (New Haven: Yale University Press, 1978).

CHAPTER EIGHT. *Humankind and the Earth*

Page 130, line 11:
Henry Thoreau, *The Maine Woods* (New York: Crowell, 1961).

NOTES

Page 130, line 24:
Lewis Mumford, quoted in Anne Chisholm, *Philosophers of the Earth* (New York: E. P. Dutton, 1972).

Page 131, line 7:
John Burroughs, *The Writings of John Burroughs* (Cambridge: The Riverside Press, 1913), vol. 15, p. 201.

Page 131, line 13:
Burroughs, *Writings,* vol. 11, p. 263.

Page 131, line 27:
Joseph Wood Krutch, *The Desert Year* (New York: William Sloane Associates, 1952).

Page 134, line 10:
Jan van Wagtendonk, "Wilderness Fire Management in Yosemite National Park," in Schofield, *Earthcare,* p. 325; H. E. Wright, Jr., "Landscape Development, Forest Fires, and Wilderness Management," *Science* 186 (1974), 487.

Page 135, line 24:
Roderick Nash, *Wilderness and the American Mind,* rev. ed. (New Haven: Yale University Press, 1973).

Page 136, line 2:
Roderick Nash, "Nature in World Development: Patterns in the Preservation of Scenic and Outdoor Recreation Resources," Lectures delivered at Bellagio Study and Conference Center, Dec. 1–5, 1976 (New York: The Rockefeller Foundation, March 1978).

Page 138, line 2:
Fred B. Eisenman, Jr., "Who Runs the Grand Canyon?," *Natural History* 87 (1978), 82; Nash, "Nature," p. 32.

Page 141, line 26:
Charles Lindbergh, *Autobiography of Values* (New York: Harcourt, Brace, Jovanovich, 1977).

NOTES

Page 144, line 6:
René Dubos, "Biological Memory, Creative Associations, and the Living Earth," in William Heidcamp, ed., *The Nature of Life* (Baltimore: University Park Press, 1978), p. 1.

Page 144, last lines:
François Jacob, "Evolution and Tinkering," *Science* 196 (1977), 1161.

Page 148, line 8:
Evon Z. Vogt and Ethel M. Albert, *People of Rimrock: A Study of Values in Five Cultures* (Cambridge: Harvard University Press, 1966).

Page 149, line 4:
David Ehrenfeld, *The Arrogance of Humanism* (New York: Oxford University Press, 1978).

Page 149, line 27:
Masaki Nakajima, "A Dream for Mankind," *PHP* (Tokyo), January 1979, p. 27.

Page 150, line 26:
Ehrenfeld, *Arrogance.*

Page 151:
Personal communication by Rufus E. Miles to the author, October 2, 1978.

Page 157, line 1:
Liberty Hyde Bailey, *The Holy Earth* (New York: Scribners, 1915).

Index

INDEX

INDEX